T0073664

THE UNIVERSAL TIMEKEEPERS

THE UNIVERSAL
TIMEKEEPERS

Reconstructing History
Atom by Atom

DAVID J. HELFAND

Columbia University Press
New York

Columbia University Press
Publishers Since 1893
New York   Chichester, West Sussex
cup.columbia.edu

Library of Congress Cataloging-in-Publication Data

Names: Helfand, D. J. (David J.), 1950– author.
Title: The universal timekeepers : reconstructing history atom by
    atom / David J. Helfand.
Description: New York : Columbia University Press, [2023] |
    Includes bibliographical references and index.
Identifiers: LCCN 2022062047 | ISBN 9780231210980 (hardback) |
    ISBN 9780231558686 (ebook)
Subjects: LCSH: Cosmochronology. | Cosmology. | Atoms.
Classification: LCC QB466.C67 H45 2023 | DDC 523.1—dc23/eng/20230609
    LC record available at https://lccn.loc.gov/2022062047

Printed in the United States of America

Cover design: Henry Sene Yee
Cover images: Shutterstock

Contents

Acknowledgments

Having recently attended my fiftieth college reunion, it is appropriate that my expressions of gratitude begin with the Amherst College 1970s physics faculty, who nominated me a half-century ago for membership in Sigma Xi, the scientific honor society. This event (coupled with my payment of annual dues) has brought me, six times a year, *The American Scientist* magazine. As I note in chapter 10, it was a particularly enchanting article in this magazine that sparked my interest in how we can reconstruct history—and prehistory—using atoms.

After a few years and a few more articles in that magazine and elsewhere, I had collected enough material to design a course for nonscience majors at Columbia University; I called it "The Universal Timekeeper." At one point, I had hoped this inherently multidisciplinary course might evolve into the Core Curriculum science course I had been trying to add to Columbia's curriculum for decades. It didn't, but that goal was eventually fulfilled by the course that inspired my last book, *A Survival Guide to the Misinformation Age: Scientific Habits of Mind*. Nonetheless, I still occasionally teach "The Universal Timekeeper" and must call out for special mention my class in the fall term of 2022 who, with the motivation of bonus points, corrected many errors in the proofs of this book, leaving few, I trust, for the reader to discover. Caroline Nicholson, in particular, was the most perspicacious proofreader by a wide margin.

I am also indebted to two anonymous reviewers who suggested that I add a glossary to this book and helped me to clarify several descriptions in the text.

A particularly fortuitous and salutary reconnection with a former student of mine from Quest University Canada, Nessa Bryce, and her sister Maggie led to one of the most exciting and fulfilling aspects of completing this book: designing the marvelous illustrations. I learned so much about how to visualize scientific concepts from those two, who effortlessly blend scientific insight, technical skill, and aesthetic sensibility into storytelling figures. If you have a scientific concept to be visualized, I highly recommend getting in touch with them immediately at BeyondBoundsCreative.com. Also, thanks to Quest graduate Graham Streich for his perspicacious reading of the manuscript.

Miranda Martin was a most helpful and cheerful editor throughout the process, and Jennifer Crewe, the director and publisher of Columbia University Press, is to be acknowledged for the most original pitch I could have imagined for me to publish this book with the press. She told me: "My father was the first person to *see* an atom." (He was, indeed: Albert Crewe, the inventor of the scanning transmission electron microscope.)

Finally, as always (or for the past forty-four years, at least), I thank my chief critic, connoisseur, and constant companion, Jada.

THE UNIVERSAL TIMEKEEPERS

Introduction

They are 99.9999999999995 percent empty space. Almost perfectly nothing.

And yet they make up everything you see, touch, smell, taste, and feel. They feed and clothe you. Their motion makes you warm (and cold). They generate your hopes and dreams and memories. They exist in splendid isolation and in highly complex assemblages. They tell time. And they can reveal secrets of the past that are otherwise unknowable.

They are atoms.

You've likely never seen an atom, even though they permeate your world. This is unsurprising when you realize that they are not just mostly emptiness, they are very, very tiny pieces of emptiness—it takes 15 million trillion of them to create a single poppy seed.

We have learned how to engage directly with these bits of near nothingness so that we can ask them intimate questions, tweak their internal states, and read their complex histories. How did you end up on that page of a medieval book of hours? When were you applied to that cave wall as part of a fallow deer's rump? What was the temperature when you fell as a snowflake onto Greenland's glacier? Where were you when the Earth coalesced from a disk of material orbiting the nascent Sun? What were you and your pals doing in the first three minutes of the universe?

To these questions the atoms, suitably tempted, are ready to reply:

From the fifteenth-century prayer book, blushingly, "I was added a bit later, in 1896."

From the cavern's wall, "It was 17,150 years ago, give or take a decade."

From a kilometer down in the Greenland Ice Sheet, "−25.5°C."

From the asteroid belt, "Far from your greedy orbit!"

From a primordial atom in your fingernail, "I was about to hook up with a cute deuterium nucleus."

Atoms are also standing by to guide us through more complex inquiries:

How did human diets change through time, and when did we develop crops that led us to abandon nomadic life? When and why did the dinosaurs suddenly disappear after 180 million years of predominance? How did life arise so quickly after Earth's formation, and why are its key molecules all left-handed? When in the history of the universe was the gold in my wedding band forged?

By probing, counting, exciting, and transforming our little atomic historians, detailed answers to these and many other questions await us. But first, we require a more formal introduction.

Atomos is Greek for "uncuttable," and in their original conception, atoms were just that—the tiniest indivisible bits of matter that could, in principle, exist. Twenty-five hundred years ago, when this idea was introduced to the West, it was purely a philosophical conceit: the world was made of a huge variety of substances, and if one took a piece of, say, wood and divided it in half, and then in half again, and then again, eventually one would reach the smallest possible particle of wood, an "atom" of wood. There was no conceivable way to test this idea, of course, but it doesn't sound completely loony.

In ancient Greece, this concept lost out to an alternative philosophy, a simpler view of matter as consisting of various proportions of just four elemental substances: earth, air, fire, and water. As a consequence, the atomic picture was largely shunned in the West for almost 2,000 years. But in the sixteenth century, after careful preservation in the Islamic world, the concept of the atom reemerged in Western thought, first rehabilitated in the church's eyes as a creation of God and then, by the eighteenth century, as a subject of empirical study.

Today, we have retained the concept of fundamental building blocks but have abandoned the notion of their indivisibility. Indeed, we know an atom's constituent parts in exquisite detail—a complex, positively charged nucleus made of protons and neutrons (themselves composed of yet more fundamental entities we call quarks), orbited by negative electrons that belong to another category of particles dubbed leptons. Indeed, even our picture of these constituents as "particles"—bits of stuff in a given spot, moving at a particular speed—has fallen away, replaced by the nonintuitive quantum chimera of particle waves.

But ignoring this complexity for the moment, we have reestablished the notion that the smallest unit of any substance is the atom (or a precisely prescribed combination of them). Furthermore, we now know that the history of the universe and all it contains is written in the particular arrangements of the fundamental atomic constituents—leptons and quarks—that comprise the building blocks of all the normal matter in the cosmos today. Given an understanding of the physical laws that govern the behavior of these particles, we can read that history, just as, given the laws of grammar and syntax, we can read historical documents humans have recorded. Although atoms are not culturally biased as human historians can be, they do have biases of their own, and we will have to be watchful for those as we cross-examine our atoms to extract their historical revelations.

Most important, however, atomic historians provide us with access to times far earlier than our oldest written records, allowing us to construct a quantitative description of what historians call prehistory. Furthermore, they reveal the story of our planet before humans arrived on the scene: a chronology of Earth's climate and the evolution of its atmosphere, the dawn of life, the birth of the solar system, and even, a little self-reflexively, a history of the atoms themselves, back to the creation of their constituent particles in the first few microseconds of the Big Bang.

As noted above, atoms are tiny—trillions can dance on the head of a pin without stepping on each other's toes, even more impressive than angels. Their internal structure is, indeed, an elaborate minuet of charged particles, and the rhythm of their dance can be used to identify them across billions of light-years of space. Remarkably, the atoms we see out there are identical to those of which we are made.

What do these tiny bits of near emptiness look like? Well, if I placed a tennis ball outside my office at the corner of 120th Street and Broadway in Manhattan to represent a hydrogen atom's nucleus (the simplest kind), its electron would be orbiting somewhere between 96th Street and 145th Street roughly two kilometers away; walking at a brisk pace, it would take you half an hour to get there. And what would you see? Most likely nothing because (1) the electron, on this scale, is far smaller than a grain of sand (at least 100,000 times smaller, actually), and (2) it would be whizzing around at about 1,350 miles per *second*—a smeared out cloud of evanescent probability that you'd be *very* unlikely to catch at all.

Nonetheless, our deep acquaintance with atoms allows us a remarkable degree of control over their behavior, which in turn makes modern life possible. For

example, atom number 55, Cesium, defines the basis of our system of time—
1 second equals exactly 9,192,631,770 oscillations of a light-wave emitted by the
transition of this atom between two of its excited states of being.[1] The GPS system in your phone relies on this by using orbiting atomic clocks—accurate to one
second in 32,000 years—to tell you where the nearest Starbucks is. Indeed, your
phone itself, made of atoms, like everything else, relies on our ability to manipulate atomic constituents precisely and reliably, time after time. The food you eat,
the medicines you take, the fuel you burn when driving your car, all work because
of our ability to control the rearrangement of atoms.

While atoms are not indestructible (and we'll see later the consequences of
both their demise and their transformation), they are impressively solid building
blocks that, under condition on Earth, tend to retain their identities indefinitely.
Thus, the atoms that make up your body *are* the atoms from the food you eat, the
water you drink, and the air your breath (50 billion trillion of them per gram).
That food in turn depends on the atoms that plants, at the base of the food chain,
absorbed from the air and sucked up through their roots. The air and soil have
their atomic compositions modified by geological and biological processes taking place on Earth's surface, most recently by the collective actions of our species.
The atoms released and rearranged by these processes were originally collected
4.567 billion years ago from an interstellar cloud of gas and dust that formed our
solar system. And yes, the atoms in that cloud originated from even earlier processes stretching back to the first moments of the Big Bang.

This book covers a whirlwind journey—a quantitative history of the universe
over the past 13.8 billion years—told through a series of tales, always with atoms
in the starring roles. As we shall see, there are ninety-four leading actors in our
drama known as the elements, and to give them their due, their names will be
capitalized throughout.

The remarkable stability of these atoms and the unique signatures each provides to the careful observer allow us to reveal the entire story in impressive
detail. We can use atoms to assign precise dates to works of human creativity,
to trace the history of agriculture and human diet, to piece together the vicis-
situdes of past climate as an aid in understanding what our future might hold,
and to reconstruct the history of our solar system and the universe itself. We will
uncover art forgeries, identify the provenance of stolen statues, and determine
the causes of death of ancient fellow humans (and what they ate for lunch the day
they died). We will deduce the Earth's temperature 100,000 years ago and relate

it to the composition of the atmosphere at that time. We will date the formation of our planet and its moon and mark the origin of life on our calendar. With our exquisite understanding of atomic structure and its many variations, we can, quite literally, reconstruct history atom by atom.

The atoms record happenings in both the organic and inorganic world. From tooth enamel to plants and tiny plankton shells, from the glassy beads spewed forth by volcanoes to rocks originating deep below Earth's crust and air in bubbles trapped within Antarctic ice, the atoms bear witness to history. They often speak the truth directly, but sometimes they have secrets in their backgrounds that we must ferret out before trusting them completely. Notwithstanding this occasional need for some careful cross-examination, from our art to our food, from changing climate to planetary catastrophes, from the origin of life to the origin of the universe, the atoms will be our faithful guides.

CHAPTER 1

Calling the Witnesses to History

The guardians had no feet. The life-size sandstone statues of two temple attendants in New York's Metropolitan Museum of Art guarded the entrance welcoming visitors to the Southeast Asian galleries for twenty years. Unfortunately, they were separated from their feet some six decades ago. These sculptures' exact provenance was unknown, although they were representative of Khmer works from the tenth century. What happened to the feet? Please call the witness Scandium.

Examination of Cambodian sandstone quarries at the atomic level ultimately allowed our attendants to resume their guard duty reunited with their feet, at the Western Gate of Jeyavarman IV's capital in Koh Ker (chapter 7).

For Europe, 1258 was a very tough year. Among the tribulations were skyrocketing food prices, famines, and pestilence among both farm animals and humans throughout the continent. Medieval chronicles describe the unusual weather in the summer of 1258; Richard of Sens, writing in 1267, characterizes it as follows:

> For so great a thickness of clouds covered the sky throughout that whole summer that hardly anyone could tell whether it was summer or autumn. The hay, drenched incessantly by strong rains that year, was unable to dry out, because it could not collect the warmth of the Sun on account of the thickness of the clouds.

What happened? From forensic analysis of atoms of wood from Bali and the composition of glass beads found under the Antarctic ice cap, Samalas, an Indonesian volcano, is revealed as the culprit (chapter 11).

A mountain hike can go badly wrong. In September 1991, two Germans hiking in the Alps near the Italian-Austrian border stumbled upon a remarkably well-preserved body sticking out of the ice at the edge of a glacier. They assumed it was a deceased hiker and notified authorities.

Their hypothesis was correct—but when Ötzi, the Iceman, was finally extracted and the date of his demise determined through atomic analysis, it was discovered that he died on a hike about 5,200 years ago.

Examination of Ötzi's partially mummified body has led to fascinating insights into the European Copper Age. His clothing, tools and other paraphernalia, the food in his stomach and intestines, his tattoos, and even his well-preserved DNA have allowed a detailed reconstruction of his times. Demanding an additional account from two cousins of Strontium, however, reveals even more about his history, identifying his birthplace and chronicling his travels before that fateful day in the Alps (chapter 10).

Planetary catastrophes don't happen only in movies. The fossil record shows that, several times in the past half a billion years, life on Earth has experienced the sudden disappearance of many (sometimes most) species of plants and animals, followed by a period in which many new species emerge to take their place. The most celebrated of these mass extinction events in popular culture is the apparently sudden demise of the dinosaurs. This discontinuity in the fossil record about 65 million years ago was first noticed in the mid-nineteenth century, and its cause has been debated ever since. We now know what happened.

Once Iridium takes the stand, we see that the dinosaurs died as the consequence of a cataclysmic collision between Earth and a wandering asteroid. Our atomic historians allow a detailed reconstruction of this cosmic catastrophe and its aftermath (chapter 12).

You are literally what you eat. Some of the atoms you ingest every day, from those in the English muffin you had for breakfast to those in the glass of wine you drank with dinner, build the structures of your body—your bones and teeth, your skin cells and neurons. Some of these atoms stay with you for a lifetime, while others are sloughed off in a matter of days. But all of them—all 3,000 trillion trillion of the atoms that are you, form a record of what you ate, as well as where and when you ate it.

By demanding an account from Carbon, Nitrogen, and other elements, we can reconstruct the history of diet and agriculture over 10,000 years and chart the spread of civilization around the world (chapter 10).

Some birthdays are noteworthy. I was born on December 7, the "day that will live in infamy" (albeit some years after the actual infamous event). When I was five, that day was a big deal to me, but today, it seems far less important. Now my wife was born on February 23, a truly important date—the day a relatively nearby star exploded, the first such event since 1604, five years before the invention of the telescope. For an astronomer who studies the remnants of exploded stars, this day marked the event of a lifetime (not to mention a guarantee I will never forget my wife's birthday again). But in the big scheme of things, such dates are not of great import.

Likewise, the exact birth date of the solar system is not known and, I suppose, one could argue it also doesn't matter much. But exploring our home planetary system's creation and subsequent evolution might provide insight into why our system looks the way it does and how common other such systems must be in the cosmos.

When Rubidium and Lead corroborate each other's stories, we not only learn the age of the Earth to a fraction of a percent but, by recruiting Aluminum as well, we can even gain some insight to what was happening in our neighborhood before the Sun formed (chapter 15).

All historians have parents. After all, the atoms that will be our guides—our witnesses to history—had to come from somewhere. Remarkably, we know the story of their antecedents in considerable detail. The first three elements were all we had when the universe reached the ripe old age of three minutes; the remaining ninety-one have been cooked inside stars and in cataclysmic explosions over the 13.8 billion years since. But we even know from whence the first three kinds of atoms emerged.

By carefully questioning the anomalously heavy Hydrogen and light Helium nuclei we see in distant galaxies, we can infer conditions in the universe back to within a millionth of a second of the beginning of time itself. These atomic historians are faithful observers indeed (chapters 16 and 17).

CHAPTER 2

Conceptualizing the Atom

From Philosophy to Science

Nothing exists except atoms and empty space; everything else is opinion.

—Democritus

Nature abhors a vacuum.

—Aristotle

Democritus and Aristotle both made many statements about the material world. To Aristotle, the world was composed of four "elements" (as he called them): earth, air, fire, and water. For Democritus, matter was constructed from an infinite variety of tiny indivisible pieces he called "atoms." But neither of these descriptions was a "model" in the sense that we use that word today. They were the product of pure thought, with no sense of urgency that such thoughts be tested against the real world. They were philosophical speculations more closely related to musings on the preoccupations of the gods, whether souls lived for eternity, and how motion endowed thoughts with being. They were, most definitively, not science.

Nonetheless, statements proclaiming such thoughts as synonymous with scientific models are common in writings about ancient philosophical traditions. The following statement is but one example: "British scientist John Dalton is familiar to us as he had developed the atomic theory in modern times. But actually, [the] Indian sage Kanada formulated the atomic theory far before."[1] Such statements irritate me no end. My primary complaint has to do with their gross misuse of the word "theory"—or, to be more generous, their use of two entirely different definitions of the word "theory" without highlighting the distinction. In the above quote from the abstract of a paper appearing in the *International Journal of Research and Analytical Reviews*,[2] the first use of "theory" is in the modern, scientific sense of the term: a falsifiable model of the material world built from the results of a set of measurements and making predictions about what future tests could be conducted to strengthen (or demolish) that model.[3] The second use of "theory" in the above quote is closer to its Greek root *theoria*, which means "contemplation or speculation." To equate Kanada's version of "atomic theory" with Dalton's is absurd.

ATOMS IN INDIA

Nonetheless, because I titled this chapter "Conceptualizing the Atom," I can devote a paragraph or two to Kanada's views, which are spelled out in the Vaisheshika Sutras. There, the Hindu sage posits the *anu* as the fundamental particle (although, for reasons I have been unable to comprehend, all subsequent commentators on these Sutras refer to the basic particle as the *parmanu*). In any event, there are four types of *anu*: earth, water, fire (or light) and air, the first two of which have mass, while the second two have none. All *anu* are spherical, indivisible, eternal, and imperceptible. They can snap together to form dyads or triads or tetrads. One knows they exist (or at least Kanada does) because earth has smell, water has taste, fire makes sight, and air is recorded by touch. Kanada's views were based on *believing* in experience.

That's "the" atomic "theory," postulated sometime between 200 BC and 600 BC—yes, it's true: we don't know when this guy lived to within 400 years! Some Hindu authors advocate for the earlier date because that would allow claims of supremacy, as it predates the Greek atomic "theory" (described below). We have a commentary on his Sutras from 200 BC so it must have been before that. The text does not mention any Buddhist philosophy, which some argue means it must have been written before approximately 430 BC when Buddha's writing became well known, although I find this argument less than compelling; I could easily have written my book about atoms without once referring to Buddha either. Unfortunately, birth dates and death dates become matters of intellectual fashion rather than a question to be answered with objective data. If we only had a single fingernail clipping from Kanada—or a single page of the bark or palm leaves on which his Sutras were likely written—we could use atomic dating to nail his birthday to within a couple of decades. More on this later.

ATOMS IN GREECE AND ROME

The "everything-started-in-ancient-Greece" version of events begins with Leucippus in the fifth century BC. Or maybe it doesn't because Diogenes Laertius, crafting biographies of Greek philosophers 700 years later, insists that Leucippus never existed. (This reliance on fallible human historians will be over as soon as we can put our atoms to work.) If Leucippus did exist (as Aristotle, among others, claims), he founded a school at Abdera in Thrace about 350 miles northeast of Athens and less than 100 miles west of the Dardanelles. There he first postulated an atomic theory of matter and took on a student who became that model's most ardent advocate: Democritus, who was born into a wealthy family in that town around 460 BC (although that date is also disputed, albeit by only a decade or so).

Democritus's life was long and eventful; some accounts purport that he lived to 109, although 90 is the more common estimate. He travelled very widely including to Egypt and to Babylon. This is interesting because, throughout his lifetime, the Achaemenid Empire (or First Persian Empire) ruled from the Balkans in Europe to the Indus Valley, including Egypt and all of Asia Minor. Indeed, Democritus's father is said to have received the Emperor Xerxes when he passed through Thrace on his way to losing the Battle of Marathon, later triumphing by slaughtering the Spartans at Thermopylae, and ultimately retreating in defeat at the Battle of Mycale on August 27, 479 BCE (interesting how we know the battle dates with such precision but can't find the right century for the philosophers). Did Democritus get the atomic model from Leucippus or from Kanada? As I read neither Sanskrit nor ancient Greek, I make a poor arbiter of this debate, and I suppose we'll never know.

What we do know, however, is that the Greek version of atoms also saw them as indivisible ("uncuttable") and moving around in empty space. Unlike Kanadian atoms, however, they have lots of surface features like hooks and eyes, and either sharp points or slippery surfaces. Solid objects have their atoms locked together, whereas in liquids, they slide over each other (see chapter 3 for the modern, analogous view). Also, unlike the Indian version (and, later, Plato's and Aristotle's models), there were not an economical four types of atoms, but an infinite number of types needed to make up the many and varied objects of the material world.

The next Western philosopher to take up the atomic model was Epicurus, born in 341 BC when Aristotle was the leading philosopher of the day. He advocated a radically materialistic view of the universe—no gods, no Platonic ideals, no souls, and no afterlife to be feared. The adjectival use of Epicurus's name today—one who indulges (perhaps overly so) in fine food and wine—suggests a hedonism that was not at all part of his philosophy.

Epicurus adopted his atoms from Democritus and had them moving about independently from the macroscopic objects they made up (as we do today). He reduced the number of kinds from Democritus's infinity to a finite number—the finite number of combinations they could make explained the finite number of substances in the world.

Epicurus's materialistic philosophy was widely adopted in the Greek and Roman worlds, reaching its apotheosis in the remarkable work *De Rerum Natura* (*On the Nature of Things*) by the Roman poet Lucretius, published around 60 BC. This stunningly modern description of many aspects of nature includes atoms, endowing them with *clinamen*, an ability to change direction, that was designed to introduce free will into his otherwise deterministic world view.

As described brilliantly in Stephen Greenblatt's book *The Swerve: How the World Became Modern*, Lucretius's poem asserts that:

The elementary particles are infinite in number but limited in shape and size. They are like the letters in an alphabet, a discrete set capable of being combined in an infinite number of sentences. And, with the seeds of things as with language, the combinations are made according to a code. As not all letters and words can be coherently combined, so too not all particles can combine with all other particles in every possible manner. Some of the seeds of things routinely and easily hook onto others, some repel and resist one another. Lucretius did not claim to know the hidden code of matter. But, he argued, it is important to grasp that there is a code and that, in principle, it could be investigated and understood by human science.[4]

As we shall see, it now has been quite fully understood, much as Lucretius imagined.

Atomism in the West remained alive and well as a counterpoint to Aristotelian philosophy through the time of Galen (second century AD), but its Epicurean connection to atheism meant it was anathema to Christians and, after the early fourth century, when Constantine made Christianity the official religion of the decaying Roman Empire, it was not seen again in Europe for 1,000 years. Lucretius's remarkable poem itself, with its empiricist and openly atheistic philosophy, was lost until its rediscovery in a German monastery in 1417 AD by Poggio Bracciolini.[5]

In Renaissance Florence, Greek and Roman philosophy found fertile ground. But the Christian backlash came before the century was out. The flourishing, secular city of Florence fell under the thrall of the Dominican friar Girolamo Savonarola in 1494, who expelled the ruling Medicis and instituted a puritanical reign, including the famous Bonfire of the Vanities in which citizens were encouraged to bring secular art, books, cosmetics, and so on, to a public burning ceremony in 1497. His Lenten sermons that year were rants against the ancient philosophers and their followers: "Listen, women. They say that this world was made of atoms, that is, those tiniest particles that fly through the air. Now laugh, women, at the studies of these learned men."[6]

While Savonarola's reign ended the following year in a personal bonfire, his excommunication of atoms from Western thought lasted another century and a half.

ATOMS IN THE ISLAMIC WORLD

Although atoms disappeared in Europe, Indian sages continued to debate the structure of matter on the smallest scale, and the blossoming of philosophy, science, and mathematics in the Islamic world helped keep the discussion alive. Whether the Islamic Kalam conception of *jawhar* ("atoms") derived from the East (India) or West (Greece) in the seventh century is unclear—quite possibly both. Later, the Islamic philosopher Avicenna (aka Ibn Sina, 980–1037AD) attacked the atomists of his day, but the substance of his argument is exemplary of how the debate remained deeply rooted in philosophy rather than physics. Avicenna admitted there might well be a smallest unit of a substance that retained the properties of that substance; that is, there could be *physically* indivisible units of matter. But he vigorously objected to the idea, held by the Kalam atomists, that these smallest units were *conceptually* indivisible as well. We now know that atoms (and molecules) do indeed represent the smallest units of matter with the characteristics of the substances they comprise, *and* that atoms can be disassembled into their constituent subatomic parts, but this is hardly the model Avicenna had in mind.

RENAISSANCE REEMERGENCE

The renaissance of the atomic model in Europe had to wait until the late Renaissance itself, when Pierre Gassendi (1592–1655), borrowing heavily from Democritus, Epicurus, and Lucretius, laid out his atomic theory of matter. Again, his basis was largely philosophical, founded more in metaphysics than in physics as we know it, but it had one key difference from the ancients: his atoms were all made by God. This removed the taint of atheism from the atomic worldview, admitting it to an age that saw the dawn of experimental science. Unlike Aristotle, Gassendi was not troubled by the notion of a vacuum that the existence of atoms implied, perhaps owing to his familiarity with the experiments of Galileo, Torricelli, and Pascal using barometers to measure air pressure (experiments he evidently replicated in 1650).[7] Gassendi's atoms, like those of Epicurus, had size, mass, and shape, and while the first two properties were confined to small ranges, there was a great variety of shapes, all, of course, bestowed by God at the Creation.

The seventeenth century saw the birth of a fundamentally new approach to creating "knowledge"—*scientia* in Latin, as in Francis Bacon's dictum from 1597, "*Nam et ipsa Scientia potestas est*": "And, thus, knowledge is power." Bacon's approach required experimentation (observing and measuring) plus inductive reasoning to create models of the natural world. This empirical approach to knowledge had precedents in the philosophical schools of Kanada (discussed above) and the Greek Stoics, and in the work of the Islamic scholar Avicenna, but the advent of explicitly testable models challenged by experiments marked a profoundly new approach to understanding nature. For example, the 2,000-year acceptance of Aristotle's deduction (from pure thought) that heavy things fall faster than those of lesser weight was demolished in 30 seconds by a single experiment: in 1586, when Simon Stevinus dropped two Lead balls, one with ten times the mass of the other, off the church tower in Delft and saw them hit the ground at the same time.[8]

THE SCIENCE OF ATOMS EMERGES

Over the first two centuries of modern science (1600–1800), progress was made in establishing empirically that every substance had a smallest unit that reflected its characteristics. Robert Boyle's experiments with gases made it clear that the four Aristotelian "elements" were not elemental at all; some substances, such as water, could be broken down into other substances—in the case of water, into Oxygen and Hydrogen—although these constituents could not be broken down further. It was these latter substances, Boyle argued, that should be called elements. In the late 1700s, Antoine Lavoisier, whose brilliant career as a chemist was cut short by a French guillotine, was the first to establish that, in chemical reactions, no mass is lost, leading to the conclusion that these reactions represented simply a rearrangement of the elements involved. He, along with Joseph Priestly, isolated Oxygen as one particularly reactive element and, by 1789, Lavoisier (with the daily assistance of his wife, Marie-Anne Paulze) had compiled a list of thirty-three elements that had not (yet) been broken down by any chemical means.[9] Some of the elements he included (e.g., light and heat) represented an incomplete knowledge of physics at the time, and others were in fact compounds of several elements that had not yet been deconstructed (e.g., baryte, which is a mineral consisting of Barium, Sulfur, and Oxygen [$BaSO_4$], and silex [SiO_4]), but Lavoisier's Hydrogen, Carbon, Nitrogen, Oxygen, Sulfur, Phosphorous, and over a dozen metals still grace our Periodic Table today.

The dawn of the nineteenth century saw the critical quantitative steps that established our modern atomic theory of matter. John Dalton determined that compounds were formed from combinations of elements in weights of fixed proportion; that is, 2 grams of Hydrogen always combined with exactly 16 grams of Oxygen to make water. This allowed him to calculate the relative weights of several of the known elements, a crucial precursor to Dmitri Ivanovich Mendeleev's development of the Periodic Table (see chapter 4).

More or less simultaneously, Lorenzo Avogadro established that equal volumes of gases (under the same pressure and temperature) contain equal numbers of atoms/molecules. Indeed, he posited a distinction between atoms (which he called "elementary molecules") and molecules made of multiple elements (a distinction Dalton had missed). By mid-century, Mendeleev organized the sixty-three elements then known (some of the earlier spurious ones having been eliminated from the list by this time) into the Periodic Table, which in turn allowed for predictions of missing elements yet to be discovered. On March 6, 1869, Mendeleev presented his paper entitled "The Dependence Between the Properties of the Atomic Weights of the Elements" to the Russian Chemical Society. Modern chemistry and the atomic model on which it was based were established.

But chemists had no more idea of the size and mass of an individual atom than did Leucippus or Lucretius; the only thing clear was that they were too small to see. And while most chemists continued their quest to discover new elements and to systematize their knowledge of the known ones, many physicists of the nineteenth century remained unpersuaded that atoms existed. The French scientist Pierre-Eugene Marcellin Berthelot, acting in his capacity as minister of foreign affairs, went so far as to forbid the teaching of atomic theory in France. As late as 1897, the Czech-born Ernst Mach stated flatly at a presentation by Ludwig Boltzmann on the latter's kinetic theory of gaseous atoms and molecules, "I don't believe that atoms exist,"[10] although it appears his objections were more philosophical than based in physics.

The first datum on the size and mass of atoms came in 1827 from a somewhat unlikely source: the Scottish botanist Robert Brown. As part of his studies on the fertilization of plants, Brown suspended pollen grains in water and observed them carefully under his microscope. He saw the grains were "particles . . . very evidently in motion." Rather than concluding this agitation was the manifestation of the "life force," he repeated the experiment, first with pollen grains that had been preserved in alcohol for eleven months (and thus were clearly quite dead) and then with rocks ground to a fine powder. The same random motion was observed in all cases. What Brown was observing was nothing less than the net effect of the random collisions of individual water molecules with the suspended particles—a few extra collisions on the left would push the particle to the right, followed by a couple of nudges from below that would push the particle upward on the microscope slide.[11]

Lucretius had actually anticipated this result and its interpretation involving atoms nearly 2,000 years earlier in *De Rerum Natura*:

Observe what happens when sunbeams are admitted into a building and shed light on its shadowy places. You will see a multitude of tiny particles mingling in a multitude of ways . . . their dancing is an actual indication of underlying movements of matter that are hidden from our sight . . . It originates with the atoms which move of themselves [i.e., spontaneously]. Then those small compound bodies that are least removed from the impetus of the atoms are set in motion by the impact of their invisible blows and in turn cannon against slightly larger bodies. So the movement mounts up from the atoms and gradually emerges to the level of our senses so that those bodies are in motion that we see in sunbeams, moved by blows that remain invisible. (verses 113–140 from Book II)[12]

It was not until 1905 that Albert Einstein quantitatively interpreted this phenomenon and calculated the size and mass of atoms. The particles Brown observed undergo roughly 100 trillion collisions per second, so a statistical approach is required. Einstein showed that, while equally likely to move right or left, the total distance a particle achieves from its starting point grows with the square root of the elapsed time (figure 2.1). He then went on from this result to calculate

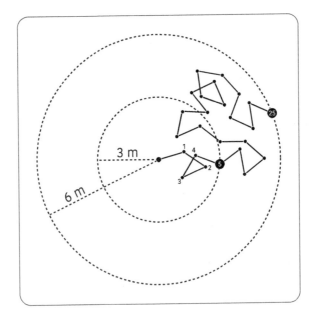

2.1 A random walk is sometimes dubbed the drunkard's walk, after an inebriated patron who steps out of a bar and falls down after each step, only to get up and take another step in a random direction. A random walk is characterized by the fact that the distance from the origin scales as the square root of the number of steps. In this case, the bar patron is 3 meters away after 5 steps and 6 meters away after $5^2 = 25$ steps.

Avogadro's number, the constant number of gas particles in a standard volume of gas that Avogadro had postulated nearly a century before. The atom had been "seen" at last.

Indeed, the last few years of the nineteenth century and the first decade of the twentieth saw rapid progress in establishing the physical properties of atoms and in disproving the "uncuttable" notion by revealing their constituent parts. In 1897, J. J. Thompson discovered electrons, showing they were both much smaller and much less massive than atoms.[13] In 1909, Ernest Rutherford and his collaborators discovered the atomic nucleus, where the positive charge and most of the mass of an atom resides.[14] This led within a few years to the Bohr model of the atom, which we will use in its modern form throughout this book (see chapter 3).[15] Meanwhile, in 1901, Max Planck introduced a new concept describing the interaction between light and matter,[16] followed by Einstein's 1905 extension of this idea to explain the photoelectric effect.[17] These developments led directly to the blossoming of quantum mechanics in the 1920s, the theory describing the behavior of the material world at the scale of atoms and smaller. A century on, it remains the most precise scientific model ever devised and provides a solid foundation on which to build our reconstruction of history, atom by atom.

CHAPTER 3

The Atom

A Utilitarian View

I t is a fundamental tenet of the scientific worldview that there exists a material reality independent both of our experiences and of our attempts to measure and interpret those experiences. Science is the process we use to construct falsifiable *models* of that reality and then test them for their fidelity to nature. This is an iterative process, and progress is more often made by proving a model wrong than by generating a profound insight de novo.

Initially, our scientific models were devised to explain (and predict) our direct experiences—the flight of a baseball; the motions of the planets; our sensations of smell and taste, heat and cold. We can touch, throw, and catch a baseball; we can see, night after night, the motion of the planets; and we can sense the aroma and taste of our coffee and note its temperature. In the case of atoms, however, we have no visceral experience to guide us—we can't see them or touch them or watch them move. But the methods of science still apply. And they allow us to build a detailed, testable, and falsifiable model with enormous predictive power, allowing us to enlist atoms in our quest to reconstruct history.

Our model here need not include everything we know about atoms and, of course, cannot include the things we don't know. But it must be a model that is fully consistent with physical reality as we know it and that captures all the atom's features essential to our project. Defining that model is the subject of this chapter.

THE HIERARCHY OF MATTER

Let us begin as any two-year-old would—with the world around us that we *can* see and touch. There appear to be thousands of different substances, each with a different color, smell, texture, reflectivity . . . many, many kinds of stuff. We have words that classify things by their use (cutlery: knife, fork, spoon), by their appearance (shiny, tarnished, clean, or dirty spoon), by the material of which they are made (Silver, stainless, or plastic spoon), and by hundreds of other categories. But if I asked you to come up with, say, just three categories—the broadest groupings possible—encompassing everything you have ever seen or felt, we might agree that there are three states of matter: solid, liquid, and gas.[1]

This classification doesn't mean we must abjure finer distinctions within those categories; the Silver spoon and the plastic spoon feel different to touch and to lift, they respond differently to heat when you put them in your cup of coffee, and they cost different amounts to replace if they accidentally end up in the garbage. But they share a common property of being *solid*—you can't compress them, and you can't change their shape (at least without considerable effort).

The coffee, on the other hand, while also incompressible, evinces a qualitative difference—it automatically takes the shape of the container into which you pour it; the *liquid* effortlessly accommodates the broad-bottomed coffee maker pot and the narrower coffee cup alike.

Finally, there is the diaphanous vapor rising from the steaming coffee. If you try to grab it, you can feel its warmth, but open your hand and nothing's there. The *gas* just dissipates.

Let us adopt the stance of Democritus for the moment and imagine that, for each of these kinds of substances, there is a smallest unit—we'll call it a "particle" for now—that retains the character of the substance. How do we imagine these smallest units interacting?

In a solid, these particles must be locked in place because when you push or pull or squeeze a solid, it maintains its shape. If you exert enough force, of course, you can effect a change—bend the Silver spoon and snap the plastic one in half—but in doing so, you have not changed the volume of the object, just its conformation. In a solid, the particles are touching each other—you can't push them closer together, and it takes a lot of effort to change even their relative orientations.

Because a liquid can't be easily compressed either,[2] its basic particles must also be touching. But there is a clear difference between a liquid and a solid in that the former can readily change its shape—indeed, it automatically does so when you transfer it from one container to another. This suggests that, while the particles are still touching, they are now free to slide over each other and assume whatever relative positions are most convenient.

Then there are gases. They are very diffuse and mostly invisible so that you often don't notice them—you don't seem to sense the air around you as you sit reading this book. But there clearly is some substance to the air; if you are out in a windstorm, you can feel the air press upon you, and you can, in some instances, smell gases (like the aroma of your coffee), suggesting something is interacting with your senses.

Indeed, gases are also made of particles but ones that are completely free of their neighbors and far from touching. It's relatively easy to compress a gas (imagine inflating your bicycle tire) because there is a lot of empty space between the particles. In the Earth's atmosphere, for example, the air particles are separated from each other by about ten times their diameters, and are free to fly around, bouncing off each other like billiard balls when they happen to meet. If you pump up your road bike tires to the recommended 116 pounds per square inch, you have just compressed the air by a factor of two in each of the three dimensions, so the particles are now separated by five times their diameters instead of ten. The pressure inside your tire is now eight times that of the air outside because the air particles are colliding with the walls of the tire $2 \times 2 \times 2$ times more often. To condense water vapor into liquid form, the particles must get $10 \times 10 \times 10 = 1,000$ times closer, which is why water is 1,000 times denser than air (approximately 10^3 kg/m^3 rather than 1 kg/m^3).

So there we have it. The three states or "phases" of matter are not distinguished by fundamental differences in the properties of the particles that make them up. It is simply a matter of whether these particles are rigidly locked in place, touching but free to slide around, or separated with lots of empty space between them. Water is water whether in its solid form (ice), its liquid form (water), or its gaseous form (vapor), and the transitions between these forms are just a matter of changing the water particles' mutual spatial relationships.

TEMPERATURE: A MEASURE OF MOTION

In discussing the transitions that a substance can make between different phases of matter, we must take a brief diversion to understand our model for heat and the metric we use to describe it, which we call temperature. As you know from everyday experience, the solid form of water (ice) is cold, and the gaseous form is hot. But what is "cold" and "hot"? It turns out these are simply the words we use to describe the relative motion of our basic particles: hot = fast, and cold = slow. What we call temperature is simply a direct measurement of the average energy of motion—dubbed "kinetic energy" (see chapter 4)—of these particles.

The basic particles of water in an ice cube are touching each other, locked in place, but vibrating (shivering, if you will) with a modest amount of energy per particle. If I raise the temperature, the particles vibrate faster. If I raise the temperature enough, the bonds that hold the particles in place will break; they are now free to slide over each other and become a liquid. This happens when I reach 32°F or 0°C. Continuing to heat the water, I make the particles move faster and faster until, at 212°F or 100°C, I liberate them entirely from their neighbors' embrace and they are free to fly off in the form of a gas.

At a given temperature, not all the particles of a substance are moving at exactly the same speed; some move faster than average, and some move more slowly. The distribution of speeds (or, more precisely, kinetic energies = $\frac{1}{2}mv^2$) is given by the curves shown in figure 3.1. Because no particle can move more slowly than zero, the distribution is slightly asymmetrical, with a few particles moving very much faster than the average (e.g., if one unsuspecting water vapor particle gets clobbered by four others all coming from the left, it will take off at high speed to the right). But overall, most of the particles have energies within a factor of two or so of the average value.[3]

The temperature scales we use to measure particle energy, like most units of measurement, are arbitrary. Zero degrees on the Fahrenheit scale was defined as such by Mr. Fahrenheit in 1724 as the coldest temperature he could get in a mixture of water, ice, and salt (records do not inform us whether he made any ice cream in this experiment). His definition of 1 degree was equally arbitrary, leading to 32° on his scale as the freezing point of pure water and 212° as its boiling point. Even these values apply only at sea level (e.g., water boils at 203° in Denver and at 190° in La Paz, Bolivia).[4] The centigrade (also called Celsius) scale was devised two decades later by a Swedish astronomer (Mr. Celsius) who decided to set his zero at the freezing point of water and count off 100 degrees to its boiling point, compared to the 180 degrees separating these two points on the Fahrenheit scale. Thus, each Celsius degree is 9/5 (180/100) of a Fahrenheit degree.

Both these scales were adopted long before we understood what "temperature" actually measured. Knowing that temperature is a measure of the average

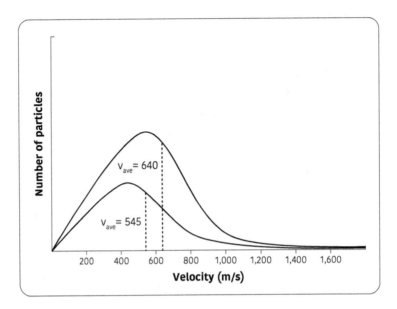

3.1 Curves representing the distribution of speeds for water particles at 0°C (32°F) and 100°C (212°F). Because no particle can move slower than 0 m/s, both curves are truncated on the left. The vertical lines represent the average speeds, which are slightly higher than the most common speed because of this truncation. Note that particles at the boiling point of water are moving with an average speed of approximately 640 m/s or 1,425 miles per hour(!), and that a few particles are moving several times faster.

kinetic energy of particles, the only sensible scale would be to set the zero point as the condition when particles stop moving ($v = 0$, so kinetic energy $= 0$). This physically meaningful scale is named after the gentleman who created our modern model for heat, Lord Kelvin. He adopted the Celsius-sized degree, so there are still 100 degrees between the freezing point and boiling point of water at sea level, but the point of zero motion has been established as −273.16°C. Thus, we say water freezes at 273 K (strictly speaking, they are called Kelvins, not degrees Kelvin) and boils at 373 K. The surface temperature of the Sun is 5,780 K, while a comfortable room temperature of 68°F = 20°C = 293 K).

This model for heat has enormous explanatory power. For example, if you are baking chocolate chip cookies, you will have probably set your oven to 375°F. If you open the door to see if the cookies are done and accidentally touch the rack on which the cookie sheet sits, you will get burned. Why? Because the particles making up the metal in the rack are vibrating at high speed, and they smack violently into your skin particles and knock them apart, turning skin particles into—well, charred skin particles. But wait, why can you stick your hand in the 375° oven? Aren't the air particles moving just as fast? Yes, they are, but as noted above, the air has 1,000 times fewer of them contacting each square inch of your skin, so while a few skin particles could be damaged, they slough off every few days anyway so no harm done. If you left your hand in the oven for 20 minutes instead of for a second or two (that is, 1,000 times longer) your hand would indeed come out looking like a burnt chocolate chip cookie. Note that you *do* feel the warmth of the oven when you stick your hand in, and that is precisely because the air particles in the oven are flying around much more rapidly than the air particles in the room, and their more energetic collisions with your skin leads you to feel "warmth."

It actually isn't true then, as I said earlier, that you don't "feel" the air around you as you sit reading this chapter. In fact, you do feel it because the temperature of the air determines the speed with which air particles bounce off your skin, which can be trillions of times per second, and that in turn engenders your feeling hot or cold or just right.

This model for heat also explains how your dishes manage to dry (i.e., evaporate all the water droplets on them) in the dish rack overnight, even though the room never (I hope) gets to the boiling point of water (212°F). On average, the speed of the water particles on the dishes is equal to the speed of the particles in the air because they are constantly colliding with each other and equalizing their energies. That average speed is not nearly high enough to liberate a water particle from the liquid state into a gas. However, recall that there *are* some water (and some air) particles moving much faster than average, and these can achieve escape speed; the droplet then loses these few speedy particles. When this happens, the average particle speed drops (if you subtract the fastest ones, the average will go down). If that were all that happened, you'd still need the dish towel in the morning. But the vast reservoir of air particles in the room still contains some of its fast particles, and when they collide with the remaining water in the droplet, the average speed goes up again and repopulates the high-speed end of the distribution (figure 3.1). These water molecules in turn can now escape, and this process continues until all the liquid water becomes a gas, allowing you to put dry dishes away in the morning.

This heat model also explains why you sweat. Your body is finely tuned to operate at a temperature of 98.6°F or 37°C, and any departure from this provokes an immediate response. If you are exercising vigorously, you turn the chemical energy stored in your muscles into heat, and the body needs to get rid of that excess heat. One mechanism for doing so is to trigger your sweat glands to make droplets of water appear on your skin. The skin particles, jiggling just a little too vigorously as far as your body is concerned, transfer some of their energy to the water particles, causing the fastest ones to escape and thus carry energy away from your skin, cooling it down. Acetone feels cold when you brush it on your skin because it has a much lower boiling point than water (only 133°F), meaning that many of these acetone particles at body temperature are going fast enough to escape into their gaseous form, carrying your skin particles' vibrational energy away with them and making the skin feel cold.

From why your bath water gets cold[5] to why the atmosphere doesn't fall down[6] (*hint:* it does), and from why your bicycle pump gets hot[7] to why your air conditioner cools your room,[8] all is explained by this simple model for heat in which temperature is simply a measure of the speeds at which particles move.

THESE "PARTICLES" ARE ATOMS AND MOLECULES

Returning to the initial theme of this chapter, we have thus far ignored what we know about the internal structure of atoms and have adopted the ancient Greek notion that each substance has a smallest unit that, throughout this discussion so far, I have been calling particles. What exactly are the particles that make up a Silver spoon or a droplet of water? Democritus and Leucippus would have argued that they are "uncuttable" (*atomos* = "atoms") and exist in an infinite variety of shapes and sizes, easily accounting for the millions of different substances that comprise our world. We know today both hypotheses are false. Atoms are far from indivisible, and those millions of substances are all composed of specific combinations of ninety-four unique building blocks.[9]

What I have been calling "particles" are either one of the ninety-four types of single blocks—for which, ignoring the etymology, we have adopted the name "atoms"—or one of millions of precisely defined combinations of atoms we call "molecules." Silver is one of the ninety-four basic building blocks and, bonded together, Silver atoms can make a silver spoon. Water is a combination of two of the basic building blocks called Hydrogen and Oxygen in the ratio of 2:1, forming the molecule of H_2O.

The Gold in your rings, the Tungsten in a light bulb filament (if you can remember incandescent bulbs), and the Silicon in the chips of your phone each represent one of the ninety-four basic building blocks we collectively call the elements (see chapter 4 for details). The breath you exhale, largely carbon dioxide (CO_2), the alcohol in your wineglass (C_2H_6O), and the sand from which that wineglass was made (SiO_2) are all molecules, atoms joined together in fixed combinations to make the myriad compounds of which our world is composed. Molecules can get complicated: the single DNA molecule that makes up human chromosome number 1 has over 13 billion atoms joined together in a precise pattern that represents part of the code for making you *you*.

DIVIDING THE INDIVISIBLE: THE BUILDING BLOCKS OF ATOMS

As noted above, atoms, despite the origin of the term, *are* cuttable. They are made of more fundamental building blocks that fall into one of two families, the leptons and the quarks. These blocks are held together by the four fundamental forces of nature that are communicated via yet another family of particles called bosons. When we catalog all the leptons and quarks we have discovered, along with their antimatter twins, and then add in the force-carrying bosons, we end up with a list of thirty-one "fundamental" particles[10]—which doesn't make them sound very fundamental at all! Many physicists believe we need to go down yet another level in the structure of matter where we may find all these different particles are actually manifestations of tiny vibrating "strings."[11] But for the moment, we're stuck with a model for nature's smallest scales comprised of thirty-one entities we will henceforth refer to as fundamental particles.

Figure 3.2 lays out the hierarchy of matter from the broad phases that opened this chapter—solid, liquid, and gas—down through molecules to atoms, then from atoms to their constituent parts. Also shown are the numbers of particle combinations that exist at each level.

Much of figure 3.2, as well as the issue about whether an even deeper layer of structure exists, is beyond the scope of our current inquiry because all we need here is a robust, accurate model of the atoms that make up the cosmos. Since the first microseconds of the universe's existence, only half a dozen of

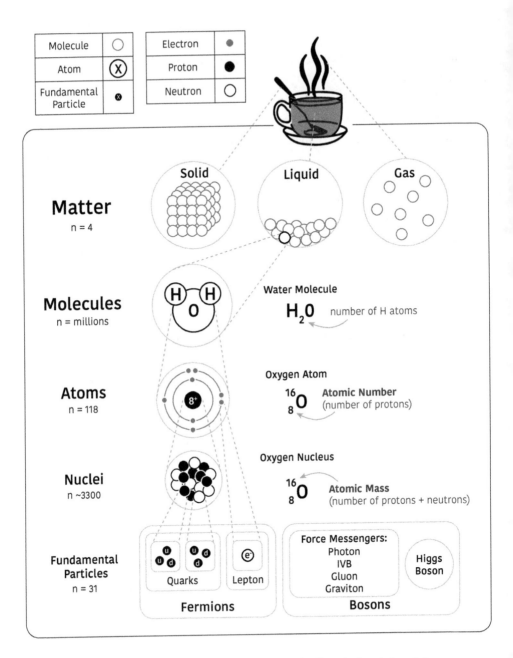

Molecule	◯
Atom	⊗
Fundamental Particle	⊗

Electron	●
Proton	⬤
Neutron	◯

Matter
n = 4

Solid

Liquid

Gas

Molecules
n = millions

(H) O (H)

Water Molecule

H_2O number of H atoms

Atoms
n = 118

8+

Oxygen Atom

$^{16}_{8}O$ **Atomic Number**
(number of protons)

Nuclei
n ~3300

Oxygen Nucleus

$^{16}_{8}O$ **Atomic Mass**
(number of protons + neutrons)

Fundamental Particles
n = 31

u u d   u u d   e⁻
d   d

Quarks   Lepton

Fermions

Force Messengers:
Photon
IVB
Gluon
Graviton

Higgs Boson

Bosons

3.2 The structure of matter, beginning with the cup of coffee, which includes solids, liquids, and gases, each made of molecules that in turn are made of specific combinations of atoms, which themselves are comprised of more fundamental particles. The number of types of particles at each level are labeled as *n*; note that the fourth form of matter (not shown in the figure) is a plasma in which the atoms are rent asunder. Only the first of the three generations of fermions are shown.

the fundamental particles are relevant: electrons (e or e^-) and their antimatter twins the positrons (e^+ or $\bar{e}$), the neutrinos associated with each of these (v_e and $\bar{v}_e$), and the "up" (u) and "down" (d) quarks, plus the force communicators. Let's address each of these in turn.

The electron, discovered in 1897, is immeasurably small (less than 10^{-18} m, a millionth of a trillionth of a meter across), with a tiny (but definitely measurable) mass of 9×10^{-31} kg. It carries a negative electric charge, which we arbitrarily set as the atomic-level unit of charge at -1. Its antimatter twin, the positron, has the same mass and size but, as with all antimatter particles, has the opposite charge, $+1$. Matter and antimatter don't mix well—indeed, if an electron meets a positron, they will mutually annihilate completely in a flash of light. All the atoms in the universe today contain electrons but, as we shall see, natural processes can produce positrons as well, and these processes will play a key role in our reconstruction of history, so we need to keep this antimatter twin in mind.

Electrons and positrons are grouped with the fermions (see figure 3.2), all of which have one additional property of importance to our story, which is fancifully called "spin." Imagine them as tiny tops that can only spin clockwise or counterclockwise with an amplitude of $\pm\frac{1}{2}$ unit (again, an arbitrary atomic unit scale on which the force-carrying bosons have a spin of 1 or 2). They interact with other particles via gravity (because they have mass), electromagnetism (because they have charge), and through the weak nuclear force (because they have a property called "lepton number" = ±1).

Neutrinos (v_e) weren't detected until 1956, although their existence was postulated decades earlier to explain the missing energy in certain nuclear reactions. They are similarly tiny in size and even smaller in mass, with an upper limit of about 1/600,000 times the mass of the electron. They are electrically neutral and also have spins with an amplitude of $\pm\frac{1}{2}$. Because they can only feel gravity (*very* faintly given their tiny masses) and the weak nuclear force, they hardly interact with normal matter at all; in the time it takes you to read this sentence, 20,000 trillion neutrinos from the Sun will pass through your body without you noticing. This is true even if you are reading this page at night when the Sun is on the other side of the Earth because they pass right through the Earth as well and will get you coming through the floor. Because they play a central role in some radioactive decays, they will be important characters in our use of atomic nuclei as clocks to chart the historical excursions to come.

In the universe today, the other family of fermions, quarks, are never seen hanging out alone; they are only bound up in pairs or triplets (see figure 3.2). The ones relevant to us are the two that combine to make protons and neutrons, the *u* and *d* quarks. First postulated in the 1960s and subsequently confirmed and described by numerous experiments at the world's atom smashers, the quarks have fractional charges: $u = +\frac{2}{3}$ and $d = -\frac{1}{3}$. Their masses are about 4.0 and 9.4 times that of the electron, respectively, and, as fellow fermions, they also have spin ±½. All quarks have an additional property unique to them—they respond to the strong nuclear force via a fourth property, which we call "color charge."

While many combinations of these and the other four flavors of quarks are possible and can be produced for fleeting moments in the lab, the two important combinations in the world today are the triplets *uud* that make a proton, and *udd* that make a neutron. A simple summation gives us the charges of these composite particles: *uud* means $+\frac{2}{3} +\frac{2}{3} - \frac{1}{3} = +1.0$ for the proton and $udd = +\frac{2}{3} - \frac{1}{3} - \frac{1}{3} = 0.0$ for the neutron. Their spins combine to yield a net value of ±½. But their masses are another matter.

It would seem obvious that the mass of three quarks would simply be the sum of the quarks' individual masses. This would yield a proton mass of 4.0 + 4.0 + 9.4 = 17.4 times the electron mass. But weighing a proton yields a very different answer: 1,836 times the electron mass, more than 100 times the simple sum (and equivalent to 1.67×10^{-27} kg). Where does all that extra mass come from? It comes from the glue that binds the quarks together. As noted above, quarks are unique among the denizens of the fundamental particle zoo because they are the only particles that respond to the strong nuclear force. This force, like its counterpart weak force, is peculiar because it exists only on scales comparable to the size of the atomic nucleus (approximately 10^{-14} m, 1 percent of a trillionth of a meter). A quark passing by a proton, say, 5 percent of a trillionth of a meter away would feel nothing.

This is radically different from the electromagnetic and gravitational forces we are familiar with in our daily lives—their ranges have no bound. The farther apart two objects with charge or mass may be, the weaker the force they feel from electromagnetism and gravity, but the force doesn't just disappear—Neptune is 2.8 billion miles from the Sun, so the gravitational attraction it feels is 900 times weaker than the force that the Earth feels, but it nonetheless keeps orbiting the Sun because of their mutual gravitational interaction. In contrast, the two nuclear forces simply disappear once one gets outside the nucleus.

So it is the gluons, themselves massless but each carrying lots of energy, that multiply the mass of the proton by a factor of approximately 100 over the summed masses of its constituent quarks (the kinetic energy of the quarks themselves, rattling around in their confining little bag, also contributes). The neutron, with a slightly heavier *d* quark replacing one of the proton's *u* quarks is slightly more massive still (by 0.14 percent). With the exception of the single proton that represents the nucleus of a Hydrogen atom, all other atoms contain both protons and neutrons bound together, as described below.

THE NUCLEUS

The core of an atom and the essence of its identity is the nucleus, a tight little ball of protons and neutrons packed together in a space only a few trillionths of a millimeter across.[12] With all those positive charges packed so closely together, the electrostatic repulsion the protons feel for each other is enormous, but the strong nuclear force wins out and keeps the particles glued together.

An atom's identity is fixed by its number of protons, and all possibilities from 1 to 94 are represented in nature. The number of protons is called the atomic number and is written symbolically as a preceding subscript for the element's chemical symbol. For Carbon, it looks like this: $_6$C. Because each element has a unique symbol *and* a unique number of protons, this nomenclature is, in a sense, redundant, so the subscript is often left off—if it is Carbon, it has six protons, and if it has six protons, it is Carbon.

The other denizens of the nucleus are the neutrons. Because they are electrically neutral, they add no electrical repulsion but contribute more of the strong-force attraction that helps to stabilize the nucleus. For the lightest couple of dozen elements, the number of neutrons and protons is usually about equal, but as we proceed up the hierarchy to heavier and heavier elements with more and more protons, extra neutrons must be added to counteract all that electrical repulsion: for Uranium, its 92 protons are usually accompanied by 146 neutrons.

Note the "usually" in the last sentence. While the number of protons unambiguously determines an element's identity, the number of neutrons is not so constrained. In chapter 5, we address the fact that a given element's nucleus can have different numbers of neutrons. Indeed, this fact is crucial for us as we employ atoms to reconstruct history.

THE ATOM

With the nucleus taken care of, we just need to add electrons to make a complete atom. This is accomplished by allowing the positively charged nucleus to attract a retinue of negatively charged electrons through the electromagnetic force: opposites attract. Because the proton and electron charges are exactly equal and opposite, a neutral atom has the same number of electrons as it has protons—Hydrogen has one, Carbon has six, Oxygen has eight, and Uranium has ninety-two.

The two forces we encounter in everyday life are gravity and electromagnetism. Gravity, the force that responds to the property of matter we call mass, keeps your feet on the ground (and your butt in your chair) and feels like a dominant force in your life—even Michael Jordan in his prime had a vertical leap under 4 feet. Electromagnetism, the force that arises from positive and negative charges, is responsible for light, for chemistry, for moving a compass needle, and for powering your cell phone, but you *feel* it only when you shuffle across a carpet, touch a metal doorknob, and get a brief shock. Experience thus suggests gravity is strong and electromagnetism is weak.

And experience is very, very wrong. In fact, the electromagnetic force is a trillion trillion trillion (yes, three of them multiplied: 10^{36}) times stronger than gravity. The key difference between these two forces is that gravity has only one sign—it is always attractive—whereas electromagnetism has two charges, positive and negative, that, in ordinary circumstances—from a single atom to a planet—exactly cancel each other. But if you had a part in a thousand (a 0.1 percent) excess of positive charges and your partner, with a 0.1 percent excess of negative charges, showed up in the doorway 10 feet away to tell you dinner was ready, the two of you would come together with a force sufficient to knock the Earth out of its orbit around the Sun. The exact equality and opposite sign of the electron and proton charges, then, is what allows a stable universe to exist.

Given this enormous attractive force between the atomic particles, one might expect that any electron wandering near an atomic nucleus would get sucked in posthaste. But that's not how things work at the atomic scale. The electron (and indeed all fundamental particles) behaves according to the laws of quantum mechanics, our highly precise model for the world of the very small. In fact, the word "particle" is not even appropriate in the macroscopic sense of this term because an electron is not a discrete, localized entity like a grain of sand; it is best described as a smeared-out cloud of probability enveloping the entire atom. In other words, the electron is not always "somewhere"—it's everywhere within its highly confined world.

While exploring the quantum world is a fascinating topic about which many books have been written, it is not a subject we need to pursue here. Our goal is to use atoms to discover art forgeries, to uncover how our ancestors learned to grow corn, to reveal what the Earth looked like covered in ice 350,000 years ago, and to learn how the solar system came into being. Thus, all I need is a *model* of an atom that accurately includes all the features needed to undertake these forensic explorations of history. That model was provided over 100 years ago by the Danish physicist Niels Bohr.

You have probably seen the iconic picture (see figure 3.3) that represents the atom as a miniature solar system, with the nucleus standing in for the Sun encircled by orbiting electrons as the planets. While this is the model we will adopt, it is important to keep in mind the limits of the solar system analogy. First, the bodies in the solar system interact through gravity, while the electrons are held in place by electromagnetic forces. Second, the spacings of the orbits are very different; while the planets orbit at distances between 41 and 3,200 times the solar diameter, the first electron orbit in Hydrogen is 24,000 times the proton diameter, a typical size ratio for atoms (recall the tennis ball at 120th Street and the electron at 145th).

Hydrogen is as simple as an atom can get—a one-proton nucleus with a single electron in orbit. Normally, that electron stays in its prescribed orbit defined above, but if the atom is struck by another particle, or if an itinerant light wave

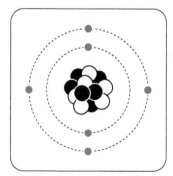

Electron	•
Proton	●
Neutron	○

3.3 Representation of the Bohr model of the Carbon atom, including the key features we will need—the protons and neutrons of the nucleus and the fixed pattern of orbiting electrons. Note the image is *not* to scale; for a nucleus the size of the image, the orbiting electrons would be roughly 500 feet away.

of the right color wanders by, the electron can be boosted to a higher orbit more distant from the nucleus. But it can't orbit anywhere it wants to, only at specific, prescribed distances (see figure 3.4). And in other atoms with more electrons, there are a set of specific distances assigned to each. This pattern of orbits is

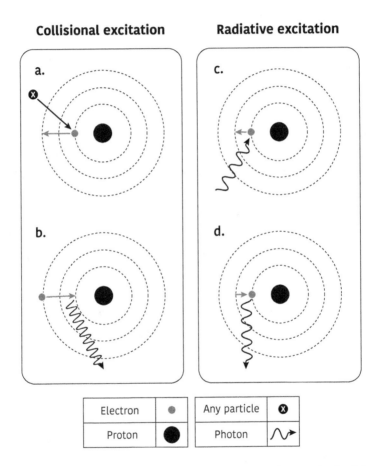

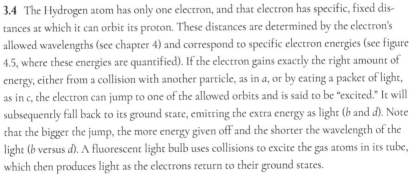

3.4 The Hydrogen atom has only one electron, and that electron has specific, fixed distances at which it can orbit its proton. These distances are determined by the electron's allowed wavelengths (see chapter 4) and correspond to specific electron energies (see figure 4.5, where these energies are quantified). If the electron gains exactly the right amount of energy, either from a collision with another particle, as in *a*, or by eating a packet of light, as in *c*, the electron can jump to one of the allowed orbits and is said to be "excited." It will subsequently fall back to its ground state, emitting the extra energy as light (*b* and *d*). Note that the bigger the jump, the more energy given off and the shorter the wavelength of the light (*b* versus *d*). A fluorescent light bulb uses collisions to excite the gas atoms in its tube, which then produces light as the electrons return to their ground states.

critical to how atoms interact both with each other and with light, allowing us to identify them across the universe; these patterns will form the subject of the next chapter, in which we explain that omnipresent wall tapestry of high school chemistry classes, the Periodic Table.

THE MOLECULE

The final step in our climb from the fundamental particles to the bits of matter that form recognizable stuff is the molecule. As described above, a molecule is a combination of two or more of the same or different kinds of atoms locked together in a fixed, specific ratio. The Oxygen you breathe in is a molecule of Oxygen with two atoms stuck together, O_2 (the subscript denotes the number of that type of atom in the molecule); the carbon dioxide (CO_2) and water vapor (H_2O) you breathe out are combinations of three atoms each (where the subscripts "$_1$" for Carbon and Oxygen are implied but not stated; see figure 3.5).

Molecules can get complicated. Vitamin C consists of only twenty connected atoms of three types—$C_6H_8O_6$—but vitamin B_{12} (otherwise known as cyanocobalamin) has 181 atoms of six types: $C_{63}H_{88}CoN_{14}O_{14}P$. Your longest DNA molecule, as noted above, contains billions of atoms.

While atoms lack the hooks and eyes and balls and sockets envisioned by the ancient Greeks, the arrangement of their outermost electrons is unique to each of the ninety-four kinds of atoms, and these electron configurations govern the connections each atom is willing to make. The next chapter explains why some are in a permanent "come hither" mode, others are on a more casual look-out for company, and yet a third group are self-satisfied loners. After joining together, the electron orbits adjust slightly, reinforcing the links, creating a specific three-dimensional shape and conferring on the molecule all its externally sensed properties like taste and smell and color and texture and density—in short, what makes water different from sand, and pomegranates different from broccoli.

While pure water (H_2O) and pure sand (SiO_2) are comprised of a single molecule each, pomegranates and broccoli contain many different molecules that determine their structure and their other properties. Indeed, many substances in everyday life are mixtures of different atoms and molecules; for example, air is a mixture of the molecules N_2, O_2, H_2O, CO_2, and a host of other less prominent ones plus the atoms Argon, Neon, Helium, Krypton, and so on. Your glass of

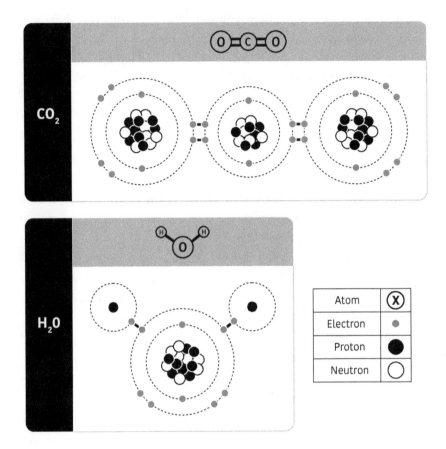

Atom	Ⓧ
Electron	•
Proton	⬤
Neutron	○

3.5 Representation of a carbon dioxide molecule where the linear arrangement of the three atoms and the double bonds between them are indicated (above) in contrast to the bent structure of the water molecule (bottom).

pinot noir is mostly H_2O, but it also includes a large collection of other molecules that confer its delectable flavor and aroma.

Take a sip of that wine as our task is now complete. Beginning with the crudest classification of the world as composed of solids, liquids, and gases, we now know these states of matter are composed of mixtures of particles, touching or not, free to move or locked in place. Matter in each of these three phases is comprised in turn of molecules, themselves constructed by taking fixed ratios of ninety-four basic building blocks called atoms. The atoms are built from specific combinations of electrons, protons, and neutrons, and the latter two entities are themselves composite bags of three quarks each.

These details are far removed from Democritus' vision of the atomic world, but the spirit is the same: the universe is composed of a set of basic building blocks that have unique intrinsic properties and combine in myriad ways to construct the richness of our reality. Adopting the Platonic term *element*[13] for these basic building blocks, we are now ready to explore the ninety-four varieties in greater detail so that they may serve as our guides to times past.

CHAPTER 4

The Elements

Our Complete Set of Blocks

Element—one of those bodies into which other bodies can decompose, and that itself is not capable of being divided into other.

—Aristotle[1]

A s evidence for the existence of atoms accumulated in the late eighteenth and early nineteenth centuries, it became clear that, unlike most of the hundreds of substances being studied, the building blocks of some kinds of matter resisted being broken down into more basic forms. Antoine Lavoisier, focusing on these irreducible substances, entitled his 1789 treatise "Elements of Chemistry" and listed a total of thirty-three elements. Some, as noted earlier, such as "light" and "caloric" (the latter thought to be a fluid that conveyed heat) represented misapprehensions about the nature of matter and energy, and others were actually molecules composed of several elements; however, twenty-three were true elements that we recognize today.

As described in chapter 3, the Bohr model of the atom includes all the features we need to enlist our atoms as inveterate historians. Their basic planetary-like structure was shown in figure 3.3. In this chapter, we elaborate on this basic model to illustrate the shells and subshells of the atom along with the rules for filling them with electrons. This will in turn both explain the structure of the Periodic Table of the Elements and set the stage for quantifying the interactions among the atoms that comprise the stuff of the universe.

THE PERIODIC TABLE

By the time Dmitri Ivanovich Mendeleev published his important systematiza-
tion of the elements in 1869, his tabulation included sixty-three substances and
predicted the existence of several more. The only three properties of the elements known at the time were their relative weights (derived from the ratios
of the masses in which they combined with other elements) and the alacrity
and selectivity with which they formed such combinations. Mendeleev's predictions of new elements to look for were based on regularities in the properties of
the known elements that grouped them in columns according to behavior, with
mass increasing down the rows. Without knowledge of the size, mass, or internal
structure of an atom, let alone any notion of what caused the behavioral patterns,
this arrangement proved prescient, and we retain it today as the Periodic Table
of the Elements (see figure 4.1).

Elements in the left-hand column are extremely reactive—Hydrogen (H)
explodes easily (the Hindenburg disaster is a notable example[2]), and Lithium
(Li), Sodium (Na), and Potassium (K) burst into flames if you drop them into
a beaker of water. They love combining with elements in the next-to-rightmost
column—which are also highly reactive—to form very stable compounds like
salt (e.g., NaCl). But the elements next door in the rightmost column can't be
induced to join with any of these neighbors, nor with any other elements in the
table.[3] To see why the various groups of atoms behave in such radically different
ways requires a quantification of the Bohr model to account for the quantum
nature of the atomic world.

As was noted in chapter 3, electrons are not like planets (or scaled-down
grains of sand). They operate in the quantum world, which means their behavior is a combination of the properties we attribute to particles *and* the properties we attribute to waves. Both particles and waves can transmit energy
(a concept we will explore in more depth below) from one place to another.
If I throw a baseball to you, it stings when you catch it because of the kinetic
energy (energy of motion) I imparted to the ball when I threw it. A particle
carries energy with it as it moves from one place to another. A wave carries
energy too, although no material need move from me to you for it to do so. If we
both hold the ends of a rope, I can jerk my end up and down, and a wave travels

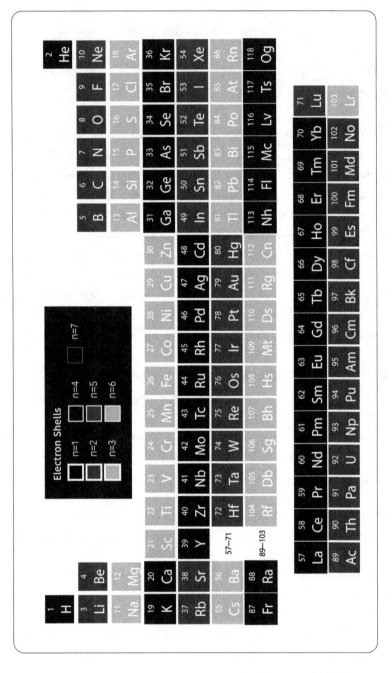

4.1 The Periodic Table of the Elements. The atomic numbers and symbols for each of the 118 types of atoms are shown. The gray scale and outline thicknesses are designed to illustrate the electron shell structures explained below. In the left-hand column, each row starts a new energy level, but the $n = 3$ and $n = 4$ levels overlap (i.e., the $n = 3$ element 18 (Ar—Argon) is followed by the $n = 4$ elements 19 (K—Potassium) and 20 (Mg—Magnesium), after which the $n = 3$ elements 21–30 intervene). The overlaps get more complicated as we move down the table, leading to the 57–71 and 89–103 strips displaced at the bottom of the Table. See figures 4.3 and 4.4 for the details.

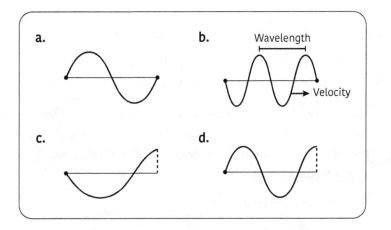

4.2 In *a*, one whole wavelength just fits between the two anchored ends of the string. In *b*, two whole waves fit in (an overtone one octave higher). But in *c* and *d*, we see that different wavelengths—slightly longer and slightly shorter—are impossible because the condition that the ends of the string remain fixed is violated.

through the rope to convey that motion to your hand while the rope particles in my hand remain there.

Any wave is characterized by two numbers—the distance between two adjacent crests (called the wavelength) and the speed with which the wave moves forward (its velocity). If we tie down both ends of a string, as in a bridge on one end and your finger on the other at a particular fret along the neck of a guitar, only certain wavelengths can possibly fit in, corresponding to the "note" you are playing (see figure 4.2). If you double the length of the string, you get a note an octave[4] lower because a wave with twice the wavelength now fits perfectly.

Stretching this analogy just a little, electrons can only *exist* at locations inside an atom where a whole number of their wavelengths fit (see box 4.1, in which this calculation is done for a Hydrogen atom). The result is that electrons can inhabit orbits at fixed distances from the atomic nucleus. This leads to basic shells labeled as $n = 1$ for the shell closest to the nucleus, $n = 2$ for the next one out, $n = 3$, and so on. As we shall see, these shells correspond to the rows of the Periodic Table.

While this basic one-wavelength-per-shell story is correct, the exact configurations of these encircling waves complicate things a bit because there is a

---

BOX 4.1 HYDROGEN ENERGY LEVELS

The wavelength of a particle in quantum mechanics is given by h/mv, where m is the particle's mass, v is its velocity, and h is Planck's constant $= 6.63 \times 10^{-34}$ J s.

The radius of the electron's orbit in Hydrogen is $r = 5.29 \times 10^{-11}$ m.
The mass of an electron $m = 9.11 \times 10^{-31}$ kg.
The velocity of the electron in its orbit $v = 2.18 \times 10^6$ m/s (about 0.7 percent the speed of light).

Thus, the electron wavelength is given by:

$$6.63 \times 10^{-34} \text{ J s} / (9.11 \times 10^{-31} \text{ kg} \times 2.18 \times 10^6 \text{ m/s}) = 3.3 \times 10^{-10} \text{ m}$$

The circumference of the electron orbit is $2\pi \times 5.29 \times 10^{-11}$ m $= 3.3 \times 10^{-10}$ m, exactly equal to the electron's quantum wavelength—the orbit is defined by one whole wave fitting around it.

The electron kinetic energy $= \frac{1}{2}\, mv^2 = \frac{1}{2} \times 9.11 \times 10^{-31}$ kg $\times (2.18 \times 10^6 \text{ m})^2$
$\quad = 2.16 \times 10^{-18}$ J
2.16×10^{-18} J $\times 1\text{eV}/1.6 \times 10^{-19}\text{J} = 13.6$ eV, the binding energy of the $n = 1$ energy level for H.

The electron in the $n = 2$ level has exactly twice the wavelength, which means the circumference of the orbit is twice as great, which in turn means the radius is $2r$. The electric force falls off as 1/distance squared, so $1/(2r)^2 = \frac{1}{4}$ times the $n = 1$ binding energy $= 13.6$ eV/4 $= 3.4$ eV.

This means a transition from $n = 2$ to $n = 1$ gives off the energy difference, 10.2 eV, as observed.

Thus, $n = 3 \Rightarrow 13.6$ eV/9 $= 1.51$ eV, $n = 4 \Rightarrow 13.6$ eV/16 $= 0.85$eV, and so on (see figure 4.5).

---

second number we are required to assign to each electron corresponding to the shape of its orbit (in physics terms, its orbital angular momentum). We label this term l, where l runs from 0 (for a spherical pattern) to 1 (three dumbbell-shaped orbits in the x, y, and z directions), to 2, 3, 4, and so on, as the orbits get more complicated. These slight differences in orbital shapes for a given shell are called subshells. Finally, as we noted in chapter 3, each electron is like a little top that either rotates clockwise or counterclockwise, a parameter we designate as spin, where $s = +\frac{1}{2}$ or $s = -\frac{1}{2}$.

The absolute rule in the quantum realm is that, while all electrons have exactly the same mass and charge, no two particles in an atom can be identical in all respects; that is, no two electrons can have the same n, l, and s numbers. Furthermore, each shell is allowed only a specific number of subshells: the $n = 1$ level has only $l = 0$ electrons; while $n = 2$ can have $l = 0$ and $l = 1$; $n = 3$ can have $l = 0$, 1, and 2, and so on. Finally, each orbital shape (l value) can host $2 \times (2l + 1)$ electrons, where the leading 2 is for one spin $+\frac{1}{2}$ and one spin $-\frac{1}{2}$ electron, and the parenthetical expression represents the number of possible orbit shapes each l-level has available. This distribution ensures that no two electrons are identical. The diagram in figure 4.3 makes all this clear.

While nineteenth-century scientists had no clue about the internal structure of atoms, they were busy discovering how light interacted with them, and they found patterns that correspond to the shell and subshell structure we use today. We maintain the arcane nomenclature these scientists concocted by assigning lowercase letters to each l subshell: $l = 0$ is the s subshell, $l = 1$ is the p subshell, and so on (see figure 4.3).

With all this nomenclature in place, we can now describe the configuration of the electrons in any atom by indicating the number of slots occupied in each shell and subshell. For example, for the element Chlorine, shown in figure 4.3 we'd say its configuration is $1s^2 2s^2 2p^6 3s^2 3p^5$—with the normal numbers indicating the energy levels n, the letters indicating the l-number designations, and the exponents indicating the number of electrons occupying each subshell. Note that all the subshells are filled to capacity, as they would be for any normal, relaxed atom, with the exception of the outermost 3p subshell, where we just run out of electrons: Chlorine has 17 protons so it must have 17 electrons and that leaves one empty slot in the 3p level. When an atom has all its electrons nestled as close to the nucleus as possible, we say it is in its "ground state."

Herein lies the key to an element's affinity (or lack thereof) for other elements: atoms like to have filled shells. Chlorine is almost there—one more electron and it would be satisfied. Even just sharing one from another atom would be better than this gaping hole on level 3p. Maybe it could borrow the lonely electron in Hydrogen and transform itself into the strong acid, hydrochloric acid:

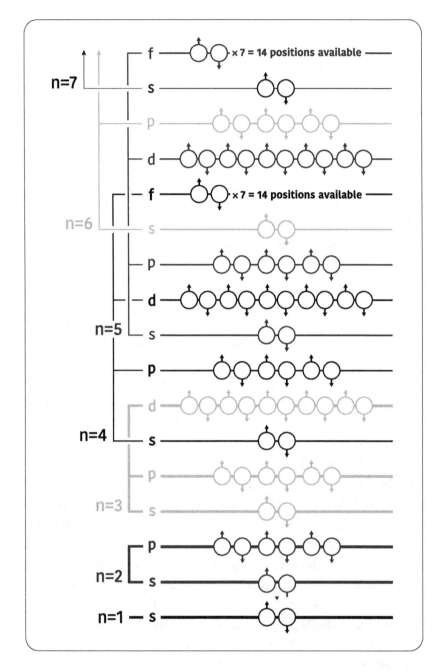

4.3 The electron shell structure for the elements in the Periodic Table. The *n* values represent the energy levels and correspond to the row number at the left-hand side of the table. Note that, starting with *n* = 3, the subshells of subsequent energy levels overlap. The s, p, d, and f sublevels correspond to the various angular momentum quantum numbers, *l*, that electrons can have; this parameter defines the shapes of their orbits (*l* = 0, 1, 2, and 3).

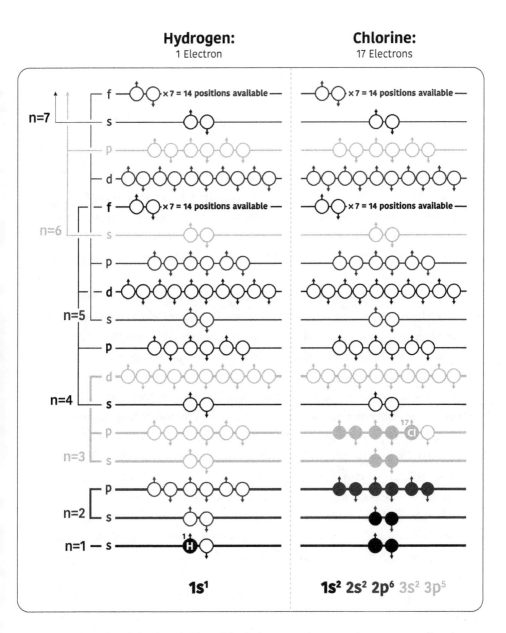

Hydrogen:
1 Electron

Chlorine:
17 Electrons

4.3 (cont) Each such level can hold $2 \times (2l + 1)$ electrons. The arrows (pointing up or down) correspond to the spin quantum numbers allowed ($s = +\frac{1}{2}$ or $-\frac{1}{2}$). No two electrons can have the same three quantum numbers n, l, and s; the circles show the possible locations consistent with this rule. Hydrogen, on the left, has just one electron, which sits in the $1s^1$ position. Chlorine, with seventeen electrons (right column), fills the two 1s, the two 2s, the six 2p, the two 3s, and five of the six 3p slots. The gray scale here, keyed to the energy quantum number n, matches that in the Periodic Table (figure 4.1).

HCl. Or it could have the almost-as-lonely outermost electron of the row three element Sodium (forming NaCl, otherwise known as table salt); or even row 4's first element Potassium to make potassium chloride, KCl, a table salt substitute for those on a restricted salt diet.

Note that the element in the table after Chlorine, Argon (Ar), has one more proton and thus one more electron, allowing it to complete its 3d shell. This makes Argon one of those snooty atoms completely indifferent to approaching suitors because its outer subshell is filled, and it has no need to share or borrow electrons to feel complete. All the elements in this rightmost column are called "noble gases" to indicate their complete unwillingness to associate with any of the rabble in the rest of the Periodic Table.

This system makes the structure of the first three rows of the Periodic Table clear. Hydrogen has an electron configuration of $1s^1$ and Helium (He) has one of $1s^2$; Hydrogen is on the prowl, and Helium is comfortably complete. The next eight elements in row 2 are Lithium ($1s^2 2s^1$—sort of desperate), Beryllium ($1s^2 2s^2$—less so), Boron ($1s^2 2s^2 p^1$—one hanging out there), Carbon ($1s^2 2s^2 2p^2$—lots of room for action with two to share and four to borrow), Nitrogen ($1s^2 2s^2 2p^3$—perfect for partnering with an identical twin, given three outermost electrons or three holes, depending how you look at it—remember this near the end of this chapter), Oxygen ($1s^2 2s^2 2p^4$—happy for a threesome with two Hydrogens), Fluorine ($1s^2 2s^2 2p^5$—yearning for a single partner), and Neon ($1s^2 2s^2 2p^6$—happy as a clam). Then we begin again on row three: Sodium ($1s^2 2s^2 2p^6 3s^1$—just like Lithium and Hydrogen), and so forth (figure 4.4). Thus, the columns of the Periodic Table contain atoms with similar outer electron configurations, each waiting to share or borrow with a similar degree of alacrity, while the rows indicate that we've moved to another orbit farther from the nucleus.

Note that in row 4, the structure of the table changes because the 4s level actually lies a bit below the 3d level, which has space for $2 \times (2 \times 2+1) = 10$ electrons (see figure 4.4), so row 4 starts with Potassium and Calcium, with outermost electrons $4s^1$ and $4s^2$, respectively, and then switches to Scandium with . . . $4s^2 3d^1$ and Titanium (. . . $4s^2 3d^2$) through Zinc (. . . $4s^2 3d^{10}$) before switching back to Gallium, with the 4p level electrons (. . . $4s^2 3d^{10} 4p^1$), and so on. Row 5 mirrors this pattern, starting with 5s electrons and then switching to fill in the 4d subshell before going back to the 5p level (see figure 4.4.).

In row 6, it gets even more complicated because the 4f level (that holds up to fourteen electrons) slips in between the 6s and 5d levels; the $6s^2$ of Barium (Ba)

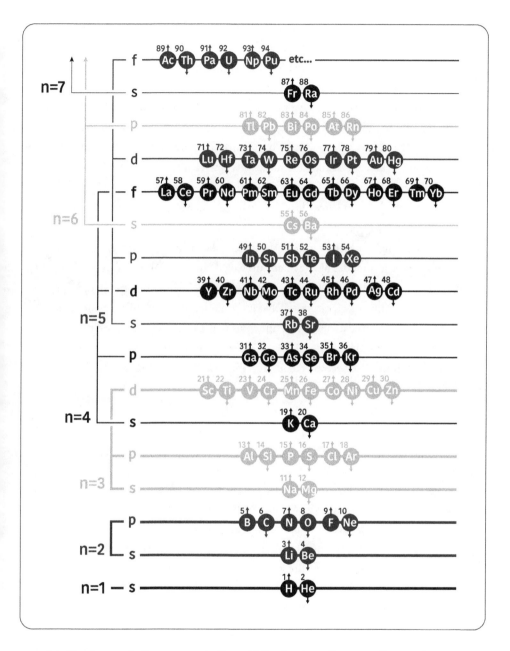

4.4 The electron shell structure as in figure 4.3 for the ninety-four naturally occurring elements. Each electron is labeled with the element symbol that represents the outermost electron for its corresponding element (the atomic numbers are also noted). For example, Aluminum (Al) has thirteen electrons, with the outermost one in slot 3p¹. The gray scale and line thicknesses match those in figures 4.1 and 4.3.

is followed by the fourteen so-called "lanthanide" elements (inserted in the table with an asterisk) before $5d^1$ Lutetium (Lu) and $5d^2$ Hafnium (Hf). This system is repeated in row 7 after element number 88 (Radium $7s^2$). Actinium's $5f^1$ electron sneaks in, following which we find the five heaviest, naturally occurring elements and another twenty-four we have managed to manufacture (albeit briefly) in the laboratory. The first five artificially produced elements live from centuries to a year, with the next nineteen living from months to mere milliseconds; unsurprisingly, none of these are available for sale at Walmart. There is a theoretical island of stability at element 126, but given that the current record holder, Oganesson, is only number 118 and lives less than two tenths of a second, getting to that island may be an impossible task.

The seemingly arbitrary rules articulated above have a solid grounding in our mathematical model of the quantum world, and they allow us to *predict* the behaviors reflected in the Periodic Table of the Elements. But the details that our understanding of quantum mechanics provides are largely irrelevant to our quest. We simply need a model that allows us to count, manipulate, and interpret the lives of atoms. The rules laid out here provide the required framework. To apply them in the service of reconstructing history, we need only understand their grounding in the concept of energy.

ENERGY

Energy is a foundational concept in the models we build of the material world, inextricably linked to atoms as we understand them. Unlike some other terms in physics, the technical definition for energy is very close to that of its use in everyday speech: energy represents the ability to do work—to push or pull; to create, resist, or change the speed or direction of motion; to transform matter from one form to another. The great utility of our concept of energy is that, while it comes in many guises and is readily changed from one form to another, it is never created or destroyed. In physics-speak, we say energy is *conserved*.

Each of the four forces of nature manifest different forms of energy. Gravity requires that objects with mass move toward each other. All the particles of the Earth attract all the other particles of the Earth, so we say the Earth has a gravitational binding energy that is equivalent to the energy required to deconstruct the Earth, particle by particle, and cast all the particles into infinity. This is the amount of energy that was released as all the planetesimals of the early solar system accreted to form the Earth. What remains of this energy is manifest today in the 6,000 K core temperature of our still-cooling planet. An object held above the surface of the Earth will fall "down" when released; that is, it has *potential* gravitational energy because it is attracted to the center of the Earth (which defines the direction "down"). Hydroelectric dams and water mills work by capturing this potential energy as the water cascades downward, generating electricity or turning the waterwheel.

The electromagnetic force manifests energy in a variety of ways. It rotates a compass needle and focuses high-energy particles from the Sun onto the North Pole to create the northern lights. Oscillating electrons in a wire power your lights and your cell phone and turn the wheels of an electric train. Light is the ultimate form of electromagnetic energy—a massless, self-reinforcing wave of oscillating electric and magnetic energy rocketing through space at 300,000 km/s (see box 4.2 for a summary of light's wavelengths, energies, and corresponding temperatures). Two particles, one positive and one negative, attract each other and, in analogy with gravity, have electrical potential energy when held apart and electrical binding energy when locked together. The energy released (or absorbed) when atoms rearrange themselves to form (or break apart) molecules—also fundamentally electromagnetic—is called chemical energy.

The nuclear forces, confined as they are to the scale of the atomic nucleus, are dominant in their domain and produce enormous binding energies that, when released, can produce high-speed particles or photons and can transform one kind of an atom to another. Even mass itself is a form of stored energy and, if released when matter and antimatter particles combine or when a new nucleus is formed, can produce a prodigious amount of energy, given by Einstein's famous equation $E = mc^2$, where m is the mass and c is the speed of light. Both the Sun and nuclear power plants are examples of mass-to-energy conversion.

As we saw in chapter 3, heat is also a form of energy represented by the motion of the particles that make up a substance. Such energy of motion, whether on microscopic (jiggling atoms) or macroscopic (a speeding passenger train) is called kinetic energy and is equal to one-half of the mass of the object in motion times its velocity squared ($E_k = \frac{1}{2}mv^2$).

While for all other physical quantities, Americans (and Bermudans) use the archaic English units (inches, feet, miles, pounds, quarts, etc.), even we have adopted the metric units for energy. The most familiar of these is the Calorie,

BOX 4.2 THE ELECTROMAGNETIC SPECTRUM

For an electromagnetic wave, energy is inversely proportional to wavelength (shorter wavelength means more rapid oscillations, which in turn means greater energy). In particular:

$E = h\,c/\lambda$ where the terms are as follows:
   E = energy of the wave packet called a photon (in joules)
   h = Planck's constant, a basic constant of nature = 6.63×10^{-34} J s
   c = the speed of light, another constant, which in empty space
      = 3×10^8 m/s
   λ = the wavelength of light (distance between two adjacent crests)

For atomic-level processes, it is somewhat more useful to quote energies in eV rather than Joules:

$$1\text{ eV} = 1.6 \times 10^{-19}\text{ J};\ 1\text{ keV} = 10^3\text{ eV};\ 1\text{ MeV} = 10^6\text{ eV}$$

Any object with a temperature greater than absolute 0 radiates electromagnetic radiation with a wavelength inversely proportional to the temperature. Higher values for T means faster particle motion, which in turn means shorter λ:

λ_{max} = 0.0029 m/T [K] where:
   λ_{max} = the peak of the spectrum, where the maximum amount of
      energy is radiated
   0.0029 m = a constant used so that the result come out in meters
   T [K] = the temperature measured in Kelvins

The spectrum is (somewhat arbitrarily) divided into unequal tranches given different names, although, in fact, it is a continuous spectrum with no limit on either end:

Name

Radio	Microwave	Infrared	Visible	Ultraviolet	X-ray	Gamma Ray

◄--+----------+----------+----------+---+----------+---------+------+-►

λ: 1 km	10 cm	0.1 mm	1um	0.1 um	0.01um	10^{-5} um	10^{-8} um
E: 10^{-9} eV	10^{-5} eV	10^{-2} eV	1 eV	10 eV	100 eV	100 keV	100 MeV
T: 3×10^{-6} K	3×10^{-2} K	30K	3,000K	30,000 K	3×10^5 K	3×10^8 K	3×10^{11} K

which one can find on the label of food products ranging from ice cream to cheese doodles. One Calorie (with a capital C[5]) is the energy required to raise the temperature of 1 kg of water by 1°C (see—metric all the way). The Calorie count on a food package, then, is a measure of the chemical energy stored in the package that will get released in the form of heat, fat cell formation, and so on, when the food in that package is digested.

The rate at which energy is used is called power, and our familiar unit of measuring power, also metric, is the watt. One watt represents the use of one joule of energy per second, where a joule is (from the kinetic energy formula of mass times velocity squared) 1 kg m²/s², or roughly the energy of a 2.2-pound chicken moving at a casual walking pace. There are 4,184 joules in one Calorie (the odd factor arising from the fact they are defined very differently—comparing the heating of water to a strolling chicken has no obvious direct equivalence).

Most of the Calories you consume are used to raise your temperature from the ambient temperature of the room (roughly 20°C or 68°F) to the temperature at which your body operates best (37°C or 98.6°F). Because your body is mostly water, we can easily calculate that this takes your mass (say, 65 kg or about 140 pounds) times 37° − 20° = 17°C or 1,100 Calories to accomplish. It turns out that a pint of Ben and Jerry's ice cream has just about 1,100 Calories (it's right there on the label) so, you might think, all you need to do is to eat one pint of Ben and Jerry's and you're good to go.

That would be true if you didn't lose any energy to the environment or need energy to keep your heart pumping the blood around your body and your neurons firing furiously as you read this sentence. In fact, to keep your body temperature in its favorite operating range in the face of the energy radiated away, as well as to maintain all your other bodily functions, you use energy at just about the same rate as a 100-watt light bulb: 100 joules/second. That means your total energy need is 100 joules/seconds × 60 seconds/minute × 60 minutes/hour × 24 hours/day = 8,640,000 joules. Converting that to Calories, we have 8,640,000 J × 1 Cal/4,184 J = 2,065 Cal per day—just about what you get in a standard diet.

The energy that keeps you ticking has had an epic journey. Originally released from a nuclear reaction in the core of the Sun hundreds of thousands of years ago, it rattled around in the Sun for millennia before breaking free from the surface, rocketing to Earth as light in a little over eight minutes, being absorbed by a plant leaf to trigger photosynthesis and form chemical bonds, being eaten by that chicken and transformed into a meaty wing, which in turn was consumed by you, and in your gut, chemical bonds were once again rearranged to generate heat to keep you warm—nuclear to electromagnetic to chemical to kinetic, the forms of energy look very different, but the quantity remains unchanged.

ELECTRON BINDING ENERGY

Now we have that pint of Ben and Jerry's tucked away and energy understood, we can turn back to the electrons encircling the atoms in those fixed patterns and understand how they interact with light, with colliding particles, and with adjacent atoms they might like to hook up with to form molecules.

Each of the electron shells and subshells defined above correspond to a particular amount of "binding energy." Because the force of electrical attraction gets weaker with distance, the electrons closest to the nucleus are the most tightly bound. We quantify this relationship by defining the binding energy as equal to the energy it would take to liberate an electron completely from the atom; such a process is called "ionization" and the resulting charged atom is an "ion." Because an electron with a binding energy of zero could logically be said to be unbound (it owes no allegiance to its former nuclear partner), we characterize the binding energies as negative; that is, we have to add energy to a negative number to get to zero.

The energy-level diagram for Hydrogen is shown in figure 4.5. The electron in the 1s state has a binding energy of -13.6 eV (see box 4.1), where eV stands for "electron-volt"; 1 eV represents a tiny amount of energy appropriate for discussing individual atoms and their constituents and is equal to 1.6×10^{-19}J. If I gave that Hydrogen electron $+13.6$ eV, it would become unbound (ionized). If I gave it 14 eV, it would use the first 13.6 eV to liberate itself and then glide away with a kinetic energy of 0.4 eV. If I gave it 25 eV, it would zip away at 2,000 km/s, arriving in Minneapolis from New York in 1 second.

Energy can be delivered to the restless electron in one of two forms. If a 14 eV photon of light zips by close enough, the electron can capture it, destroying it completely and transforming its electromagnetic energy into the kinetic energy necessary for its escape. As an alternative, another atom, a molecule, or a subatomic particle like another electron, can collide with the atom; if its kinetic energy is greater than 13.6 eV, again, the electron can escape.

Any photon or colliding particle with less than 13.6 eV of energy will likely just pass the atom by or bounce off without changing anything. While the

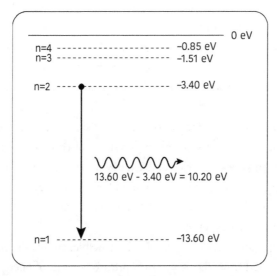

4.5 The energy-level diagram for the Hydrogen atom, showing the binding energies of the various *n*-levels (the subshells have been suppressed for simplicity). The electron can absorb energy from a colliding particle or a packet of light if either has the right amount of energy to boost the electron to one of the allowed upper levels. When the electron falls back down, it either produces light or gives the extra energy to a colliding particle in an amount equal to the difference in energy between the levels. The level labeled 0 eV corresponds to the electron becoming unbound from the nucleus, the process we call ionization.

Hydrogen atom only has one electron in its 1s shell, the higher-level shells still exist, and if the illuminating photon or colliding particle have just the right amount of energy, the electron can get "excited" by jumping from the ground state to one of the higher levels. For example, if a photon of exactly 10.2 eV zips by, it is likely to be captured and destroyed because that's exactly the amount of energy the electron needs to jump to the 2s level, leaving the 1s level temporarily vacant. For a colliding particle, its kinetic energy only needs to be greater than 10.2 eV because the electron can absorb the amount it needs, and the incoming particle can bounce off, carrying away the rest. Outward bound, that particle will be moving more slowly because it has given up part of its energy to the electron, so we call this an inelastic collision (the energy coming in is not equal to energy going out because the electron gobbled up some of it; see figure 3.4).

The electron in the 2s level of Hydrogen is now said to be "excited." Being a somewhat laconic particle, it is not likely to stay excited very long. Left alone, it will jump back to the 1s state in about 0.125 seconds, on average. This jump down produces energy, and that energy can be carried away in the form of a 10.2 eV photon—an excited atom can create light. It is also light of a very specific energy unique to the separation of energy levels in each kind of atom, thus allowing us to identify Hydrogen and Helium and Carbon atoms as such clear across the universe.

An alternative route to deexcitation is a collision. In a dense environment, an atom can be struck by its neighbors hundreds of times per second, and any random bump can knock the electron back down to the 1s state. In this case, that 10.2 eV of energy still needs to be accounted for, and it ends up in the colliding particle, producing a super-elastic collision in which the outgoing particle gains energy at the electron's expense.

The Hydrogen atom has many levels (in principle, an infinite number), with each level separated from the others by a precise amount of energy. Transitions up and down between any pair of levels is possible (although some are more probable than others, and the lifetimes in each excited level vary greatly), leading to a wide variety of energies that photons and colliding particles can have to trigger an electron jump.

In more complex atoms with more electrons, even more transitions are possible. The inner levels of high atomic number atoms such as Uranium are very tightly bound because the dozens of protons in the nucleus all attract the closest electrons, leading to binding energies of more than tens of thousands of eV. The outermost electrons of most atoms are held more or less as tightly as the Hydrogen electron because, to the outermost electron, the atom sort of looks like Hydrogen; all the electrons closer to the nucleus cancel one positive charge each, so the lonely outside electron effectively sees just one positive charge. In Uranium, for example, the innermost electrons have a binding energy of 115,000 eV, and the outermost one (in the 6d level) has a binding energy 16.8 eV, just 24 percent greater than Hydrogen's sole electron.

NUCLEAR BINDING ENERGY

In chapter 3, we saw how, in the atomic nucleus where all four of the forces of nature are in play, the attraction of the strong nuclear force overcomes the electrical repulsion of all the tightly packed, positively charged protons to create the heart of the atom. We can determine the stability of an atomic nucleus by measuring the strength of this attractive strong force and subtracting the repulsive electrical force to find the net energy with which the particles are attached to each other: the nuclear binding energy. This represents the amount of energy required if you wanted to pull all the protons and neutrons apart and scatter them to the four winds. Equivalently, given conservation of energy, it also represents the amount of energy given off when the particles snap together to form the nucleus. This is completely analogous to the electron binding energy described above, but the strength of the strong force and the much smaller space that the nucleus occupies means that nuclear binding energies are much larger: instead of the 1 to 100,000 eV range we saw for electrons, nuclear binding ranges from 1 million to 9 million eV for each nuclear particle. The total binding energy of a Carbon nucleus with six protons and six neutrons is 92.1 million electron volts (MeV), whereas its six electrons have a total binding energy of 632 eV—the difference in energy between my strolling chicken and a Harley-Davidson going 75 miles per hour.

The electron volt is indeed a small unit of energy, so even 92 million of them is not a lot on a human scale. But atoms are small too, and if you add the amount of nuclear binding energy in the Carbon atoms in a flake of graphite from your pencil, it's roughly the energy of a 1.5 million pound, six-car passenger train traveling 80 miles an hour! And that's why nuclear transformations—be it in a power plant or a bomb—are so much more potent than chemical reactions, where each atom's interaction with its neighbors yields roughly 10,000,000 times less energy.

We can calculate the binding energy of any nucleus by simply weighing it (or, more precisely, measuring its mass). You might think that such a measurement wouldn't even be necessary because we know the number of protons and neutrons in each nucleus, so we can just sum the masses of its constituent particles to find the total. But as Albert Einstein taught us, mass is just another form of energy, and all that energy binding the nucleus together has to come from somewhere. In fact, that somewhere is mass: $E = \Delta mc^2$, where E is the binding energy of the nucleus, Δm is the difference between the sum of the masses of the constituent particles and the mass of the nucleus itself (c is the speed of light).

We arbitrarily pick the Carbon atom to define the unit of atomic mass (cleverly called the "atomic mass unit," abbreviated amu). With the scale set so that a Carbon atom has a mass of 12.000 amu, we can do the calculation in box 4.3 to find the binding energy of a Carbon nucleus. Summing the atom's constituents,

---

BOX 4.3

On the Carbon = 12.000 amu scale, a single proton has a mass of 1.00728 amu, and the neutron has a mass of 1.00867 amu. Thus, six protons weigh

$$6 \times 1.00728 \text{ amu} = 6.04368 \text{ amu worth of protons}$$

and six neutrons are just a bit heavier at

$$6 \times 1.00867 \text{ amu} = 6.05202 \text{ amu worth of neutrons}$$

We can't forget the electrons. They have a very small but nonzero mass of 9.1×10^{-31} kg each, so six of them add another

$$6 \times 0.000548 \text{ amu} = 0.00329 \text{ amu}$$

meaning that the total atom contains 12.0989 amu of parts. Subtracting 12.000 from this and converting to energy units, we get

$$0.0989 \text{ amu} \times 1.66054 \times 10^{-27} \text{ kg/amu} \times (2.99792 \times 10^8 \text{ m/s})^2 = 1.476 \times 10^{-11}$$
$$\text{joules}/1.6022 \times 10^{-13} \text{ J/Mev} = 92.1 \text{ MeV}$$

---

we find a total mass of a little over 0.8 percent above 12.000 amu. When we bring all these constituents together, however, that extra mass is given off as energy—as we shall see in chapter 16. This is why the stars shine. Converting mass to energy using Einstein's equation, we get 92.1 million electron volts of binding energy for the Carbon atom—the speeding bullet train in a flake of graphite.

CHEMICAL BINDING ENERGY

Our final topic in this chapter is understanding the process by which molecules are formed. By process of elimination, it is clear that the force tying atoms together into molecules must be electromagnetism—gravity is vastly too weak to be important on the atomic scale, and the two nuclear forces don't extend beyond the nucleus. Yet how do the interacting electrons—all negatively charged and thus mutually repulsive—form strong connections between atoms?

The answer lies in the effective distribution of the electron waves around the nucleus, plus the desire of the atoms for the symmetry that comes from full electron shells. The Hydrogen and Oxygen atoms in water provide a good example. Each Hydrogen atom with just one electron would be happier if its 1s shell were either full (with two electrons) or empty (with no electrons). The Oxygen atom has the configuration $1s^2\ 2s^2\ 2p^4$ and would like to fill its two remaining 2p slots. This provides a match made in atomic heaven. The Hydrogen atoms can each share their single electrons with the Oxygen atom so its 2p level feels fuller. Meanwhile, with the electrons hanging out more around the Oxygen atom, it becomes slightly negatively charged, on average, while both Hydrogens become slightly positively charged. And positive and negative charges attract (see figure 3.5).

This asymmetric charge distribution in the water molecule—called polarity—accounts for many of water's important properties. First, it means water molecules attract each other (positive end to negative end to positive end . . .) producing what we call surface tension—like a skin on a puddle, strong enough that some insects can literally walk on water. It also explains water's ability to dissolve almost anything; the electrical forces tug apart the weak bonds that hold the molecules of other substances to each other. This is the feature of water that makes it so central to life—it dissolves all kinds of chemicals and transports them around as it flows through a plant stem or a blood vessel. Polarity also explains the unusual property of water that its solid phase is less dense than its liquid phase (i.e., ice floats).

To make water, one literally "burns" (combines with Oxygen gas, O_2) Hydrogen gas (H_2) through the reaction represented as:

$$2H_2 + O_2 \rightarrow 2H_2O + 19.2\ eV$$

That is, every two H_2 molecules (so four Hydrogen atoms in total) connect with one O_2 molecule (two Oxygen atoms) to make two molecules of water (two H_2O) and, in the process, they give off 19.2 eV of energy. This means, on average, each O-H bond in a water molecule has a binding energy of $19.2eV/4 = 4.8\ eV$. This is typical of bond energies in relatively simple molecules, which generally lie in the range of 1 to 10eV. One of the strongest common bonds is found in the Nitrogen molecules that make up most of our atmosphere—the N_2 bond takes 9.8 eV to break. Because Nitrogen is an essential element for plant life, this poses a problem. Plants can't break apart the N_2 in the air and must relegate this task to bacteria that live on their roots; they use Oxygen as an energy source to disrupt the N_2 bond, making the Nitrogen atoms available for the plants to use (see chapter 10).

As noted earlier in this chapter, the asymmetric distribution of charge in a molecule allows molecules to be attracted to each other. In water, this is a relatively strong attraction, amounting to 0.42 eV per molecule in the liquid state, meaning this is how much energy has to be added to break the connections between the water molecules in the transition from the liquid phase to a gas (otherwise known as boiling water to make steam). To go from the solid to the liquid phase of water (i.e., melting ice) just means partially breaking the connections to allow the molecules to slide over one another, and it requires only one seventh as much energy (0.06 eV per molecule).

We have now come full circle from the beginning of chapter 3 to the end of this one, chapter 4. We started grouping everything into three phases of matter and saw it was simply the strength of the connections between the basic particles that determined their state. We now see that changing a solid to a liquid and a liquid to a gas takes roughly 0.05 eV and 0.5 eV, respectively. Breaking the particles into their constituent atoms takes roughly 5 eV per bond. To break atoms into electrons and nuclei takes 15 to 150,000 eV, breaking nuclei into protons and neutrons takes 5–10 million eV, and breaking protons into quarks takes 50 million eV. The energy scale spans a factor of 1 billion, but the principles remain the same: from quarks and leptons to ice cubes, a hierarchy of structure, governed by the four forces of nature, links the fundamental particles into the ninety-four basic building blocks that shape our world. While now known not to be "uncuttable," these atoms have become familiar friends whose inner lives and external relations are well understood. After chapters 5 and 6 present a brief exploration of their familial relationships and expected lifespans, we will be ready to join them on a journey through history.

CHAPTER 5

Isotopes

Elemental Flavors

H aving delineated the unique status of the ninety-four kinds of atoms that make up all the millions of substances comprising our world, it is time to introduce a slight complication: not all Carbon atoms are the same after all, nor are all Hydrogen atoms or Oxygen atoms or Uranium atoms. Each of the ninety-four elements comes in two or more flavors that we call "isotopes."

Iso in Greek means "the same," and *topos* means "place."[1] All Carbon atoms *are* in "the same place" in the Periodic Table, which, you will recall, means they all have exactly the same number and arrangement of electrons and thus behave the same way in all chemical reactions. And, as we know, if they have the same number of electrons, they must also have the same number of protons to remain electrically neutral. Indeed, this is the case—the atomic number of every Carbon atom in the universe is six, indicating its six protons and six electrons.

Thus, the only option left for distinguishing multiple flavors of Carbon is the number of neutrons each nucleus contains because, being neutral, these particles leave the electrically charged components unaffected. And it is indeed in the number of neutrons in each atomic nucleus where we see distinctions. If you took a fingernail clipping and deconstructed it atom by atom, you'd find about 45 percent of the atoms were Carbon. But if you sorted the atoms very carefully by mass into tiny buckets, you'd end up with three different buckets of Carbon. About 98.93 percent of the Carbon atoms would be in a bucket with an atomic mass of 12 amu. Most of the remaining 1.07 percent would have a mass of 13 amu, and roughly one in a trillion atoms would weigh in at 14 amu. You might think that anything present at only one part in a trillion would be unlikely to show up at all, but it is important to remember that atoms are *tiny*;

from your one fingernail clipping, there would be more than a billion atoms in the bucket labeled 14.

The different isotopes of Carbon are thus distinguished by the number of neutrons their nuclei contain, and the outward manifestation of these differences is the atom's mass (originally dubbed "atomic weight"). We indicate these differences by labeling the atomic symbol with the mass number (simply the number of protons plus the number of neutrons) as a preceding superscript: for the three most common isotopes of Carbon, we have ^{12}C, ^{13}C, ^{14}C (sometimes also written as C-12, C-13, and C-14).

In total, there are another dozen isotopes of Carbon ranging from ^{8}C (with only two neutrons) to ^{22}C (sporting a whopping sixteen neutrons) but none of these, whether made in the lab or in nature, stick around very long; ^{11}C has an average lifetime of 20 minutes or so, and all the rest live less than 20 seconds (some much less, such as ^{8}C with a lifetime of 0.000000000000000000002 or 2×10^{-21} seconds!). ^{14}C is also prone to falling apart, but it does so at a leisurely rate measured in thousands of years. The propensity for some nuclei to transform themselves spontaneously into another isotope is the basis of radioactivity, the subject of the next chapter and the key to dating events long past.

Thus, we say that Carbon has two *stable isotopes* (^{12}C and ^{13}C) and thirteen unstable or radioactive ones. This doesn't mean that the stable isotopes are completely incapable of change. If we whack their nuclei hard enough, either with other particles or with extremely high-energy photons, we can both excite them and even transform them into other nuclei. But left in peace, they will live as themselves, untransformed for at least a billion times the age of the universe— very stable.[2]

THE DISCOVERY OF ISOTOPES

Just as chemists in the first decades of the nineteenth century introduced the concept of atoms with different masses and characteristics, physicists in the first two decades of the twentieth century discovered that an elemental atom could exist in different mass states. Two independent threads came together by 1920 to establishment the existence of isotopes.

The first line of investigation involved the radioactive[3] elements at the end of the Periodic Table, Thorium and Uranium. Uranium is derived from the mineral pitchblende, a substance used as a colorant in glassmaking since the time of the Roman Empire. It was first isolated in elemental form in 1789 by the German apothecary Martin Klaproth, who named it after William Herschel's first telescopically discovered planet Uranus found earlier in the decade. The Swedish chemist J. J. Berzalius identified the new element Thorium in 1828, one of eight different elements[4] discovered in his laboratory during the first three decades of the nineteenth century.

The serendipitous discovery of Uranium's radioactivity by Henri Becquerel in 1896 (see chapter 6), followed quickly by the identification of Thorium as a companion radioactive species in 1898 (by Gerhard Schmidt and, independently, Marie Curie), intensified study of these two elements. It quickly became clear that ores containing them also housed other radioactive substances, tentatively named mesothorium (from Thorium ores) and ionium (from Uranium ores). Attempts to isolate these two potentially new elements by chemical means found a surprise, however: ionium could not be distinguished from Thorium itself, and mesothorium appeared chemically identical to Radium, two steps down the Periodic Table. The English chemist Frederick Soddy summed up the situation in 1910: "elements of different atomic weights may possess identical [chemical] properties."[5] In other words, atoms with identical electron configurations (which defines their chemical properties) and thus identical numbers of protons (which defines their place in the Periodic Table) can have different masses ("atomic weights").

Given that relative atomic weights were a key factor in identifying the proportion of each type of atom in compound substances—and thus were important in identifying the elements themselves—this conclusion was somewhat disconcerting. A final resolution of the matter awaited the arrival of a new technology: the mass spectrometer. On December 1, 1919, Francis Aston, a physicist working in the Cavendish Laboratory at the University of Cambridge, published a paper describing his "positive ray spectrograph."[6] This device used a combination of electric and magnetic forces to deflect the "positive rays" (what we now call ions) emitted by various substances into discrete buckets based on their charge-to-mass ratio. Working with Neon gas, Aston showed that a stream in which all the ions had the same charge divided into two separate buckets with atomic weights of 20 and 22. In the following years, he used his device to identify 212 distinct, naturally occurring isotopes from dozens of different elements.

Aston found in his experiments that, when isotopes of the same element were separated from each other, they each had an atomic weight very close to a whole number on the scale, discussed in the last chapter, in which Carbon was exactly 12 units and Hydrogen was 1. Thus, for example, naturally occurring Chlorine, long measured as having an atomic weight of 35.45 times the weight of Hydrogen, was actually a mixture of two different isotopes of the element: 75.77 percent Cl-35 and 24.23 percent Cl-37.[7] With the discovery of the neutron in 1932 by James Chadwick, the meaning of Aston's "whole number rule" became clear: the atomic weight (or, today, atomic mass) is simply the sum of the number of protons and neutrons that a nucleus contains. Different isotopes are distinguished by the number of neutrons accompanying the fixed number of protons (the atomic number) that defines each element's place in the Periodic Table (figure 4.1).

ISOTOPY INVENTORY

With only two exceptions, every element from number 1 (Hydrogen) to number 82 (Lead) has at least one stable isotope. Twenty-six elements have only one (e.g., Beryllium, element number 4; Fluorine, element number 9; Sodium, number 11; Aluminum, number 13; etc.[8]); the record holder is Tin (element number 50) with ten stable isotopes. These eighty elements have among them a total of 254 isotopes that have never been observed to be unstable in any way.

Among the first eighty-two elements, the two lacking a stable form are Technetium (atomic number 43) and Promethium (number 61). The two longest-lived isotopes of Technetium are ^{97}Te and ^{98}Te, both averaging about 4.2 million years, although the most common isotope found (along with Uranium in the mineral uraninite[9]) is ^{99}Te, which is present in the minute concentration of about one part in 4 trillion but only lives 211,000 years. This means Technetium can't possibly be left over from when the Earth was born 4.57 billion years ago, so these isotopes must be created through natural processes involving the decay of heavier elements, processes we will discuss in the chapter 6. Likewise, Promethium is found in Uranium-bearing minerals in tiny quantities; the most stable isotope is ^{145}Pm, with an average lifetime of only 17.7 years.

Beyond Lead (number 82) in the Periodic Table, none of the naturally occurring elements (numbers 83–94) has even one stable isotope, although some are very long-lived and have been around since the formation of the solar system. Bismuth-209, element number 83, is the record holder, with an estimated lifetime of 1.9×10^{19} years, or more than 1 billion times the age of the universe—it's not going anywhere soon. The other two long-lived isotopes from this part of the Periodic Table are Thorium-232, with a lifetime of 14 billion years (within a few percentage points of the age of the universe) and Uranium-238, at 4.47 billion years, almost exactly the age of the Earth. As we shall see in chapter 6, "lifetime" here is not an absolute number. Just as average life expectancy for humans doesn't mean everyone dies at 78.6 years old—plenty of older people exist—so too we would expect some of each of these three elements to have been on Earth since its formation. Including the radioactive isotopes of elements that have both stable and unstable forms, there are total of thirty-four different isotopes that are unstable but have lifetimes of greater than 100 million years. These unstable but long-lived isotopes are fittingly called "primordial" because they were present in the cloud from which the solar system condensed.

The other nine heaviest elements live much shorter time spans—from 80.8 million years for Plutonium-244 (number 94) to only 22 minutes for Francium-223 (number 87). Even the dozens of extra neutrons attempting to hold these nuclei together can't overcome the enormous electrical repulsion of all those protons squeezed into a tiny space. These (and other) isotopes found in nature are not from the Earth's natal matter but are constantly being formed by the decay of the long-lived radioactive isotopes of other elements. Again, including the species of all elements that are products of ongoing nuclear decay, there are fifty-three such isotopes found on Earth.

This brings the total number of naturally occurring isotopes among the ninety-four elements to 339. The adjectival phrase "naturally occurring" here is slightly misleading, however, because it refers to the very limited domain of nature comprised of our planet Earth. In the nuclear furnaces found in the cores of massive stars, in the violent explosions that end these stars' lives, and in other colossally energetic cosmic events such as the merger of two neutron stars, many more isotope varieties are undoubtedly produced (see chapter 16). Because all these varieties have short lifetimes compared to the age of the Earth, however, none of them is present in crustal minerals.

ARTIFICIAL ISOTOPES

Many of the substances of the modern world, of course, do not exist in nature—we manufacture them by combining elements to make novel molecules, ranging from polyethylene bags for our groceries to chlorofluorocarbons to run our air conditioners to streptomycin to kill bacteria. These products all come about through *chemical* interactions, the rearrangement and linking of atoms in specific proportions through their mutual electron interactions. The energies involved in these reactions are measured in electron volts (eV) per molecule (chapter 4). Might it be possible, by ramping up the energies by a factor of 10 million or so, to transform elements from one to another or even to create entirely new isotopes?

This transmutation of elemental forms was one of the ultimate dreams of *alchemy*, a widespread practice found in China, India, Europe and the Arab world in the prescientific era. The word comes to us from Medieval Arabic *al* ("the") *kimiya*, with the latter word derived from the Greek *khemia*, literally the "art of transmuting metals." The first records of this practice are attributed, ironically enough, to an author writing under the pen name Democritus (known by historians as "Pseudo-Democritus") from Hellenistic Egypt in the first century AD. The goals of alchemy ranged far beyond the standard trope of turning Lead into Gold and included searches for the potion of immortality and a substance to cure all diseases (*panaceas*, from the Greek *pan* ["all"] and *akos* ["remedy"]). Needless to say, lacking the technology to generate the millionfold increase in energy over chemical reactions required for elemental transmutation, alchemy never succeeded.

Early in 1934, however, less than two years after the discovery of the neutron completed our picture of the atomic nucleus, Frederick and Irene Joliot-Curie (Marie Curie's daughter) created the first "artificial" isotopes by bombarding stable isotopes with fast-moving Helium nuclei and creating hitherto unknown isotopes of Nitrogen, Phosphorous, and Silicon. This modern realization of the alchemists' dreams was recognized the following year with the award of the Nobel Prize in Chemistry. Over the last ninety years, not only new isotopes but twenty-four entirely new elements have been produced, and now the modern Periodic Table includes, in addition to the ninety-four naturally occurring atoms, the manufactured elements 95 through 118. The total isotope count has climbed past 3,330, of which 620 have lifetimes of an hour or greater. Some of these have found important uses in medicine, energy production, and other technical applications.

ISOTOPES OF INTEREST

To a nuclear physicist, unsurprisingly, all isotopes are interesting. But several are of wider interest because of their technological applications or, as we shall see in subsequent chapters, their usefulness in reconstructing history. Because all living matter is based on Carbon, the three naturally occurring Carbon isotopes are of great utility in a host of fields—from discovering fraud in the sale of old wine to dating old structures or works of art, and from revealing ancient diets to recording the history of climate in prehistoric times. The heavy, stable isotopes of Hydrogen (^{2}H) and Oxygen (^{18}O) are similarly useful for providing records of past temperatures and precipitation rates. In addition, ^{18}O is the key precursor for Fluorine-18, which is used in medical positron emission tomography (PET) scans.

A number of radioactive isotopes are used in medicine, both as diagnostic tools and for treatments. In addition to ^{18}F, Nitrogen-13, Cobalt-60, Galium-67, Technetium-99, Palladium-103, Ruthenium-106, Indium-111, Iodine-123, Iodine-125, and Iodine-131, Cesium-137, Iridium-192, and Thallium-201, administered by injection or ingestion, have all been employed in imaging studies, while the noble gases Krypton-81 and Xenon-133 are used with inhalers. Strontium-89, Yttrium-90, Iodine-131, Samarium-153, and Lutetium-177 have been applied to attack certain cancers and for palliative treatment of bone pain arising from a variety of diseases. Some of these isotopes have average lifetimes of mere minutes and so must be produced on-site and used immediately before they decay into other isotopes that are not useful for medical purposes.

Nuclear reactors are powered either by Uranium or Plutonium, although a new generation of Thorium reactors is now being developed. Naturally occurring Uranium, dominated by the long-lived isotope ^{238}U, usually needs to be "enriched" to boost the fraction of the shorter-lived, lighter ^{235}U to between 3 percent and 5 percent, depending on the type of reactor[10]. Inside reactors, some ^{238}U is turned into ^{239}Pu, which also contributes to the heat produced, while ^{238}Pu is used as the source of power for deep space missions that travel too far from the Sun for solar panels to be practical. While there are several potential advantages to using Thorium-232 as a nuclear fuel, including its much greater abundance on Earth, its reduced nuclear waste, and the difficulty of weaponizing its by-products, no commercial Thorium reactor has yet been built.

Isotopes have even touched the world of geopolitics. In 1958, Louise Reiss and her husband Eric helped found the Greater St Louis Citizens' Committee for Nuclear Information out of concern for the effects of the U.S. and Soviet atmospheric nuclear weapons tests, then occurring monthly. In collaboration with the dentistry schools at Washington University and the University of St. Louis, they collected over 300,000 baby teeth from young St. Louis residents. The teeth were analyzed for the presence of the radioactive isotope Strontium-90, known to be produced in nuclear explosions. Strontium lies directly under Calcium in the Periodic Table and thus has a similar outer electron distribution that allows it, when ingested, to substitute easily for Calcium in bones and teeth.

Strontium has an average lifetime of 28.8 years, meaning that more than three quarters of this isotope, once incorporated into teeth and bones, will undergo radioactive decay over the course of a human lifetime (with the accompanying, unwelcome release of high-energy radiation inside the body). In 1963, H. L. Rosenthal, J. E. Gilter, and J. T. Bird[11] were able to show that children born in 1957 had ten times as much ^{90}Sr in their teeth as those born in 1951; by 1963, the increase was fiftyfold. These data were released as President John F. Kennedy was in negotiations with the Soviet Union to ban atmospheric tests of nuclear weapons, and the results are widely believed to have led to the rapid ratification of the treaty by the Senate just seven weeks after its signing in Moscow on August 5, 1963, one day before the eighteenth anniversary of the atomic bombing of Hiroshima, Japan.

CHAPTER 6

Radioactivity

The Imperturbable Clock

In chapter 5, I introduced the notion of "radioactive" isotopes and talked about their "average lifetimes." Because these concepts are central to our goal of using atoms to reconstruct history, it is important to elaborate on and refine these concepts here.

THE DISCOVERY OF RADIOACTIVITY

"Radioactivity" has nothing to do with radio waves but derives from the Latin *radiatio* ("shining," related to the root *radius* as in the spokes of a wheel that "radiate" from the hub) and *actif* (old French from the Latin *activus*, "doing"). The word was coined by Marie and Pierre Curie in 1898 as they explored the surprising discovery of Henri Becquerel, from two years earlier, that Uranium (and, they soon discovered, Thorium and Polonium) emitted highly energetic rays of . . . something.

In 1892, Becquerel had followed his grandfather and his father to become the third generation of Becquerels to occupy the chair in physics at the Paris Museum of Natural History (in fact, four generations of Becquerels occupied the chair continuously from 1838 to 1948). Henri's primary interest was the phenomenon of phosphorescence that can be found in products such as glow-in-the-dark frisbees and those stars kids have pasted on their bedroom ceilings. Phosphorescence occurs when shining a light on an object energizes it so that it glows (often with a different color light) for a long period after the illumination ceases. In our atomic language, we now understand this as the illuminating light causing electrons to jump to excited states, where they reside for a relatively long interval before jumping down to a different energy level (and thus emitting a different wavelength of light).

Following up on Wilhelm Roentgen's surprising discovery of penetrating, invisible "X-rays" in November 1895, Henri Becquerel postulated that these X-rays might be related to phosphorescence and guessed that substances he had been working with, such as Uranium salts, might emit these same X-rays if they were illuminated by bright sunlight. In late February 1896, he prepared an experiment by wrapping a glass photographic plate in layers of black paper (to prevent direct light exposure), laying the salt sample on top of a Maltese cross placed on the paper, and exposing the salt to bright sunlight for several hours. When the plate was developed, it showed the image of the cross, consistent with his phosphorescence theory—the sunlight stimulated the Uranium salts to emit rays that exposed the film except where they were intercepted and blocked by the metal cross.

Subsequently, a few cloudy days upset this hypothesis and led to a momentous discovery. In Becquerel's words:

> Among the preceding experiments, some had been prepared on Wednesday the 26th and Thursday the 27th of February, and since the sun was out only intermittently on these days, I kept the apparatuses prepared and returned the cases to the darkness of a bureau drawer, leaving in place the crusts of the Uranium salt. Since the sun did not come out in the following days, I developed the photographic plates on the 1st of March, expecting to find the images very weak. Instead the silhouettes appeared with great intensity.[1]

An illuminating light source was completely unnecessary—the Uranium salts spontaneously emitted energetic rays that passed through the layers of black paper and exposed the photographic plates, displaying the silhouette of the cross even when secured in a dark drawer!

Following the announcement of this discovery, Pierre and Marie Curie, Ernest Rutherford, and others quickly found that there were three different types of rays emerging from radioactive materials, which they named after the first three letters of the Greek alphabet: alpha, beta, and gamma. By 1900, all three were identified.

Alpha rays were in fact Helium nuclei (which we now know means two protons and two neutrons bound together by the strong nuclear force). Beta rays were simply high-speed electrons, the negatively charged particles that had been discovered three years earlier by J. J. Thompson. And gamma rays were just very high energy (short wavelength) photons of light. Because this was still more than a decade before Rutherford's discovery of the atomic nucleus, the origin of these different "rays"—and the source of their very high energies—was a total mystery. Today we understand radioactivity as a process by which an unstable nucleus undergoes one of three types of spontaneous transformations that move it toward greater stability.

THE QUEST FOR STABILITY

As noted in chapter 3, the atomic nucleus is a battlefield between the electrostatic repulsion of the many positively charged protons packed together in a tiny space and the stronger glue of the strong force—felt by both protons and neutrons—that overcomes this repulsion and binds the nuclear particles together. For each element, defined by its proton number, there is an optimal number of buffering neutrons to make the nucleus stable (although, as noted in chapter 5, there are more than a dozen elements for which no permanently stable nucleus exists). The optimal proton/neutron ratio is displayed graphically in figure 6.1, where we indicate the "valley of stability" that steadily departs from a ratio of 1:1 as the atomic (proton) number increases and extra neutron glue is needed to keep the nuclei from exploding.

If a nucleus finds itself outside this valley of stability, it looks for a way to approach the valley and roll downhill to an energetically more comfortable place. The seven different routes it can take are illustrated in figure 6.2. These processes are referred to as radioactive "decay," although there is nothing any more dead or putrefying about the end point than there is about the original nucleus; indeed, the end product is closer to a peaceful life of stability. "Transformation" would probably have been a better word choice than "decay," but we will stick with the standard language here for consistency. Likewise, we will adopt the standard terms of "parent" for the original radioactive nucleus and "daughter" for its decay product.

ALPHA DECAY

When a heavy nucleus lies below the line of stability (which curves upward) on the graph, it can get closer to stability by moving diagonally down toward

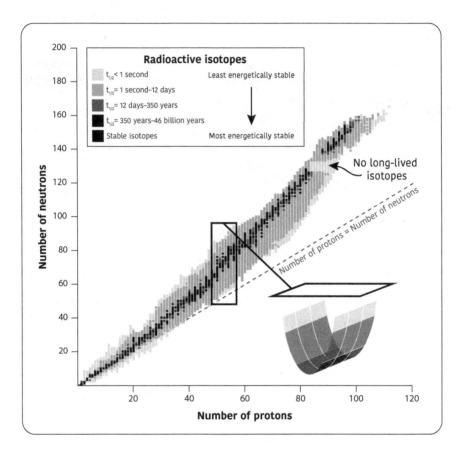

6.1 A plot of all known isotopes of the 118 elements displaying each nucleus's number of protons plotted against its number of neutrons. Most elements have one or more isotopes that are stable—their neutron/proton ratio is optimal. Most also have one or more radioactive (unstable) isotopes where the neutron/proton ratio is far from the valley of stability (represented by the darkest colors; see the inset in the figure) and an adjustment through radioactive decay is inevitable. For the lightest dozen or so elements, n = p is optimal; for heavier elements, extra padding by a larger number of neutrons is required for stability.

lower atomic numbers. This transformation is what alpha decay accomplishes; the nucleus spits out two protons and two neutrons—an "alpha particle" (aka a Helium nucleus)—and jumps two boxes down in the Periodic Table; for example, $^{238}_{92}U \rightarrow \,^{234}_{90}Th + \,^{4}_{2}\alpha$. Note that the atomic mass (number of protons plus number of neutrons) is the same on both sides of the equation (212 = 208 + 4), as are the number of positive charges (92 = 90 + 2). All radioactive decays must conserve these two numbers; neither charge nor mass number can be created or destroyed.

Energy must always be conserved as well. The reaction described above produces 4.3 million electron volts (MeV) of energy as the nucleus goes from a more tentatively bound state to a more tightly bound state—as it, metaphorically speaking, rolls downhill toward the valley of stability. As shown in chapter 4, this energy comes at the expense of the mass of the resulting two nuclei, the sum of which is slightly less than that of the original nucleus (by the Δm in $E = \Delta mc^2$ or, in this case, 0.46 percent).

Alpha decay is most relevant for heavy nuclei with isotopes lying far below the stability line. Starting with elements just above the number 50, such as $_{52}$Te (Tellurium) and $_{53}$I (Iodine), isotopes far from stability have an increasing likelihood of undergoing alpha decay; for example, with ^{127}I as the first stable isotope, ^{113}I, ^{112}I, ^{111}I, ^{110}I, and ^{108}I (with fourteen to nineteen too few neutrons) have 0.00003 percent, 0.0012 percent, 0.09 percent, 17 percent, and 90 percent chances, respectively, of decaying through the alpha channel. The farther from stability the isotopes are, the more advantageous it is to jump down two spots in the Periodic Table all at once. This trend holds through the last element with any stable isotopes, Lead, where the first significant amount of alpha decay occurs in the isotope with fourteen fewer neutrons than the lightest stable isotope ^{204}Pb.

Starting with atom number 83, Bismuth, the lightest element with no stable isotopes, alpha decay is even more common; for example, ^{209}Bi, with the remarkable average lifetime more than a billion times the age of the universe, itself undergoes alpha decay—just *very* infrequently. The same is true for all but the three shortest-lived elements among the dozen heaviest elements without stable isotopes (Astatine, Francium, and Actinium)—the longer-lived forms all undergo alpha decay.

There is one important exception to the general principle that alpha decay is confined to the domain of the heavier elements. Beryllium, element number 4, is also the one exception to the rule that the light elements are most stable when they have an equal number of protons and neutrons: ^{9}Be, not ^{8}Be, is the stable isotope. In fact, ^{8}Be undergoes alpha decay, splitting into two ^{4}He nuclei. This turns out to be critical to the way stars produce energy and live out their lives. After spending billions of years fusing Hydrogen to make Helium, and once a star's core is mostly Helium, the Helium nuclei can't fuse to make ^{8}Be because such nuclei immediately (in 0.000000082 seconds, which is pretty immediately) fall apart again through alpha decay. This has profound effects on the senility of stars and the creation of the elements themselves, both of which we will explore in chapter 16.

BETA DECAY

The recognition that beta particles were simply electrons was not terribly surprising in the context of the atomic model of 1900. That model pictured the atom as a "plum pudding" composed of positively charged pudding dough embedded with negatively charged electrons as the raisins (or plums, if you pre-fer), making the atom neutral and yet incorporating the only subatomic particle known at the time, the electron. All that was needed was some internal process that could eject these raisin-electrons at high speed. But the idea of beta decay *should* be disturbing to *you* if you have followed my story this far: in our modern atomic model there *are* no electrons in the nucleus! And remember, it is *nuclear* transformation we are talking about in radioactive decay.

Again, it is Einstein's famous equation $E = mc^2$ to the rescue. Mass and energy are interchangeable, so if you have enough energy, you can make some mass. Given the tiny mass of the electron, it is not even that hard to make one; in energy units, the electron mass is just 0.511 MeV and, as we have seen, the nuclear binding energies are measured in tens to thousands of MeV. There are some rules that still need to be followed, however. As noted above, we must conserve charge and mass number and energy.

And there is one more rule: just as we are required to preserve the number of protons plus neutrons, we must keep constant the number of leptons, the class of particles to which electrons and neutrinos belong (see chapter 3). This task is made easier by the existence of antimatter particles which, by definition, "cancel" normal particles. Thus, you can create an electron plus an anti-electron (a positron) with no problem if you have 2 × 0.511 MeV or 1.022 MeV of energy to spare—we've made one negative plus one positive charge, so they cancel, and one lepton and one anti-lepton, which also cancel (and we haven't changed the number of protons plus neutrons at all). As long as mass number and charge are carefully balanced, we can create one electron and one antineutrino, which would also take care of the lepton number balance.

It turns out it is the latter process (and its inverse—creating one positron and one normal neutrino) that characterizes beta decay. In fact, this process can take three different routes, all in the interest of moving the atomic nucleus closer to the happy valley of stability. If a nucleus has too many neutrons, as ^{12}B does, it can fix things nicely by spitting out an electron (plus an antineutrino to keep the lepton balance). What this does is effectively transform a neutron into a proton and an electron. With the extra proton, the nucleus has taken one step up in the Periodic Table, so Boron becomes Carbon. At the same time, it has added a neutron, taking the ratio of neutrons to protons from 7:5 (too many neutrons) to 6:6 (just right).

In fact, this process doesn't only happen inside a nucleus like ^{12}B. Any unlucky neutron that finds itself alone outside the embrace of the nuclear strong force will undergo the decay $_0n \rightarrow {}_1p + {}_{-1}e + \bar{v}$ in an average of 880 seconds (about 15 minutes). Inside a nucleus, the timescale for this to occur varies over an enormous range—for ^{12}B, it is 0.02 seconds, while for ^{14}C, it is 5,730 years.

This branch of beta decay—when a parent sheds an electron to reduce the neutron/proton ratio—moves an isotope from above the line of stability down and to the right, closer to that line. In contrast, isotopes below the line need to move up and to the left to roll into the valley of stability and, as a consequence, the opposite process is required, sometimes called inverse beta decay. For example, ^{12}N with too few neutrons (a 5:7 ratio) can seek stability by shooting out a positron, effectively turning a proton into a neutron: $^{12}N \rightarrow {}^{12}C + e^+ + v$ yields an ever-happy Carbon with a 6:6 ratio and all the conservation laws (charge, mass number, energy, and lepton number) obeyed. Radioactive isotopes throughout the entire Periodic Table utilize beta decay and inverse beta decay to reach more stable forms, the choice between the two being determined simply by the side of the valley they lie on.

A third route for beta decay—electron capture—also produces a more balanced nucleus. If an electron in the swarming cloud of electrons wanders too close to the nucleus, it can get dragged in, transforming a proton into a neutron like inverse beta decay and thus moving an isotope lying below the line closer to stability. The example we use here is Beryllium-7. This nucleus has too many protons and not enough neutrons, so it is thus prone to capturing an orbiting electron: $^7Be + e^- \rightarrow {}^7Li + v$, bumping itself down one step and creating the more comfortable ratio of neutrons to protons of 4:3.

GAMMA DECAY

The final kind of "radiation" emitted by radioactive nuclei, the release of gamma rays, is in fact the only one of the three that actually conforms to our standard use of the term "radiation" because it is simply high-energy light. It arises in a manner exactly analogous to the way in which electrons emit lower-energy light—a transition from an excited state to a more relaxed state. Recall from chapter 4 that when an electron in orbit about the nucleus either eats a photon or is whacked by an incoming particle, it can absorb that energy and jump to an excited state. After some time (maybe very little time indeed), it can jump back down and emit a photon of its own. The atomic nucleus has analogous excited energy levels (see figure 6.2) which can be accessed by the nucleus either through absorbing the right energy photon, undergoing a collision with an outside particle, *or* undergoing an alpha or beta decay that leaves the nucleus in an excited state. Because all the energies involved in the nucleus are millions of times greater than those that bind the electrons in their orbits, instead of photons of a few eV (visible light), we get photons of millions of eV (MeV) in the gamma ray part of the spectrum.

SPONTANEOUS AND INDUCED FISSION

The most dramatic form of nuclear transformation doesn't shift a nucleus up or down one or two steps in the Periodic Table but relocates it dramatically by splitting into two or more pieces. This process only occurs naturally for the isotopes ^{232}Thorium, ^{235}Uranium, ^{238}Uranium, ^{239}Plutonium, and ^{240}Plutonum, and even in these cases, it is extremely rare. For ^{238}U, for example, it occurs only 0.000054 percent of the time, with standard alpha decay being the normal first step toward the valley of stability. This decay mode is much more common, however, in the humanmade elements lying above Plutonium in the Periodic Table. For example, the isotope ^{250}Cm of the element number 96 Curium undergoes spontaneous fission roughly 74 percent of the time as an attractive alternative to the alpha (18 percent) and beta (8 percent) decay routes.

When spontaneous fission occurs, the nucleus never breaks into exactly equal pieces but can produce a wide variety of different elements toward the middle

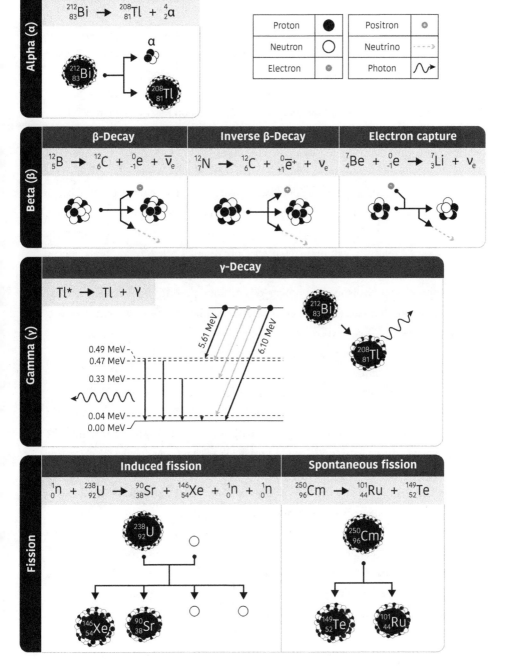

6.2 Schematic representations of the seven types of nuclear decay: alpha decay, beta decay, inverse beta decay, electron capture, gamma decay, induced fission, and spontaneous fission. The heavy nuclei are labeled with their atomic mass, atomic number, and chemical symbol. For the light nuclei involved in beta decay, the numbers of protons and neutrons are shown explicitly. The decay equations are given above each illustrated reaction.

part of the Periodic Table. In addition, however, one or two neutrons often end up without a home in either fragment, and these lead to the final of the seven decay modes: induced fission. Being neutral, neutrons have no trouble penetrating an atomic nucleus, and when they find themselves inside a heavy unstable nucleus, chaos can ensue. Most neutron-induced fission reactions yield two large pieces plus some wayward neutrons, although somewhat less than 1 percent of the time, three separate pieces are produced.

Fission can also be induced by a sufficiently high energy photon blasting the nucleus apart or when the nucleus is struck by a high-energy particle other than a neutron. But it is relatively slow-moving neutrons that are most effective. Because more than one neutron is typically produced in each fission reaction, these excess neutrons can in turn trigger more fissions, releasing yet more energy and more neutrons. This can lead to a self-sustaining reaction. When controlled by carefully monitoring the number of neutrons produced, we have a nuclear power plant that can generate electricity with one ten millionth the fuel volume of processes that involve chemical reactions such as burning coal, oil, or gas. When the reactions are allowed to multiply without limit, one gets the Hiroshima atomic bomb.

As noted above, a Uranium nucleus undergoing fission (either ^{238}U or ^{235}U) typically splits into two unequal pieces. The lower-mass isotopes are centered roughly around an atomic mass of 95 with a range from 80 to 110, while the higher-mass piece is typically around mass 135 with a range from 125 to 155 (see figure 6.3). Because these two fragments arise from a parent nucleus that is neutron rich (e.g., ^{238}U has a neutron/proton ratio of 146:92), both daughter isotopes have too many neutrons and lie above line of stability (see figure 6.1). Thus, the products of a fission reaction are themselves radioactive, typically undergoing a series of beta decays to get closer to the valley of stability. Strontium-90, mentioned in chapter 5, is an example of a radioactive fission product. Some of these species are long-lived, creating the politically intractable radioactive waste issue that haunts nuclear power generation.[2]

One other mode of nuclear transformation is the opposite of fission and is responsible for the creation of all the elements beyond the primordial Hydrogen and Helium: nuclear fusion. We will defer a discussion of that process until chapter 16, where we will describe the creation of the elements themselves in in the cores of massive stars.

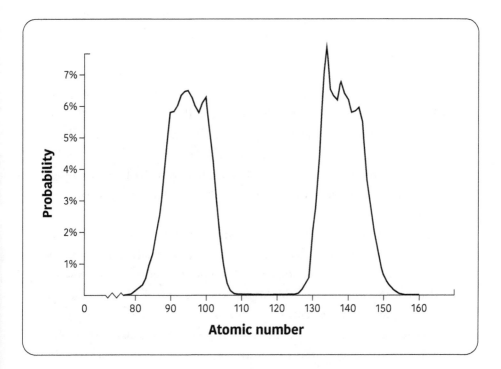

6.3 A fission reaction (induced or spontaneous) can split a heavy nucleus in many ways. The curves represent the frequency with which various mass fragments are emitted by a Uranium-235 fission. The lower-mass isotopes are centered roughly around an atomic mass of 95, with a range from 80 to 110, while the higher-mass piece is typically around mass 135, with a range from 125 to 155.

LIFETIMES AND HALF-LIVES

Throughout chapter 5, I referred to "average lifetimes" for various radioactive isotopes. Because the rates of decay provide the crucial clocks for our reconstruction of history, it is important to define much more carefully what I meant by "average lifetime." To do so, we must first recognize a central fact about radioactive decay: it is fundamentally a probabilistic process.

What this means is that, if presented with a single, radioactive nucleus, there is nothing I can do to predict when it will decay. It could decay in the next second or not for a million years. And my ignorance here is not a consequence of a lack of the appropriate tools or a failure to record the history of this particular nucleus. It is a fundamental ignorance that is an essential feature of any truly random process.

If I flip a (fair) coin, there is no way I can determine whether it will come up heads or tails. This is a random process. If on the first flip, I get heads, I still have absolutely zero knowledge of what the outcome of the next flip will be. Indeed, if I flip five heads in a row, the sixth flip has exactly a 50 percent chance of being heads and a 50 percent chance of being tails. This latter result may seem hard to accept—if I get five heads in a row, surely you'd feel my odds of getting tails on the next flip must be better than 50:50. This feeling is known as the gamblers' fallacy and provides the primary source of income for casinos all over the world: because it *is not* true. In a random process with two possible, equally likely outcomes, each realization must have exactly a 50 percent chance of one outcome and a 50 percent chance of the other. Even if I flipped twenty heads in a row—in less than one in a million tries will this happen—the odds of a head on the next flip are still precisely 50 percent.

Radioactive decay is likewise a truly random process, with a 50 percent chance for each of two outcomes: either the nucleus decays, or it does not. There is nothing I can do—even in principle—to predict when the decay will occur. Like the coin, the history of the nucleus is irrelevant. If you find this counterintuitive, you are in good company. Einstein never accepted that a natural process could be completely random and felt that our inability to predict a decay was the result of an incompleteness in our model for how the microworld works. He spent the last thirty years of his life searching for an alternative model and never found one. Our model—quantum mechanics—is now nearly 100 years old and has made by far the most precise predictions that any model of the material world has ever achieved. It has passed every test we have posed with flying colors. Einstein still could be right, of course, and a deeper layer of knowledge could reveal the inner secrets of radioactive decay, but for now, the experimental verdict is clear: decay is fundamentally probabilistic.

Just because the process is random does not mean there is nothing we can say about possible outcomes, however. If I have 100 coins and flip them all simultaneously, I can be confident in my prediction that roughly half will be heads and the other half tails. This prediction won't be exact—it won't be precisely fifty heads and fifty tails very often, but the result is much more likely to be between, say, forty-five and fifty-five heads than between seventy-five and eighty-five heads. And the odds of 100 heads? About one in 1,000,000,000,000,000,000,00 0,000,000,000—if all 8 billion people on Earth flipped coins 100 times a minute, it would take, on average, 200 times the age of the universe to flip enough times for 100 heads to happen.

Likewise, if I have a pile of identical radioactive nuclei on my desk, I can't predict what any one of them will do, but I can say with some confidence how long it will be before half of them decay (i.e., come up tails). The time interval over which half of a sample will decay, different for every radioactive isotope but easily measured for most, is called the isotope's half-life. Let's consider what this quantity means for a sample of radioactive atoms.

Suppose that, when I start my clock at noon, I have 10,000 atoms of a radio-active isotope sitting on my lab bench. If that isotope's half-life is one hour, at 1 P.M., I'll have roughly 5,000 remaining. It is very unlikely to be exactly 5,000— this is a random process after all, just as it would be unlikely to have exactly fifty heads out of 100 coin flips. But it will be roughly 5,000.

Those 5,000 nuclei have neither a memory nor a sense of time, and they certainly don't know when I started watching them. So, as is true of *all* nuclei of this isotope, each has a 50:50 chance of decaying in one hour. Thus at 2 p.m., there will be about 2,500 left; half of the 5,000 that were there at 1 p.m. will decay in that next hour. At 3 p.m., we'll be down to 1,250, and at 4 p.m., there will be roughly 625.

The number of atoms left at any time decays exponentially, as shown in the curve in figure 6.4. It is exactly like flipping coins. The odds of one head are 50 percent. The odds of two in a row are 25 percent because all four of the following possibilities are equally likely: HH, HT, TH, and TT, and only one of the four satisfies the outcome we seek (two heads). The odds of three heads in a row lead to eight possible outcomes: HHH, HHT HTH, HTT, TTT, TTH, THT, THH, and only one of the eight is the outcome we are looking for (three heads in a row). The rule in probability is that, for independent events (like coin flipping or radioactive decay), we calculate the odds of event 1 AND event 2 AND event 3 all happening by simply multiplying the probabilities of each event together. In this simple case, where the probabilities are all ½, the probability of getting *n* heads in a row is $P(n) = (½)^n$.

In the case of the radioactive nuclei, the logic is the same. Just imagine every nucleus in your sample flips a coin once per half-life. In this case, half will get tails and will decay, while the other half get heads and live on. After another

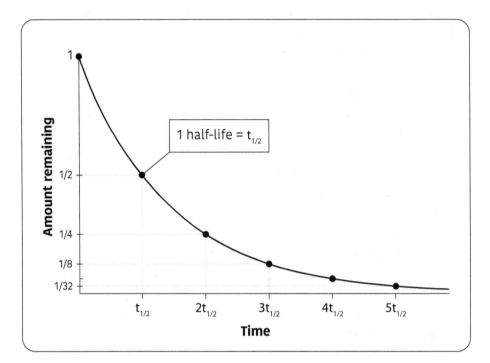

6.4 The exponential decay of a radioactive source. One half-life is the time it takes 50 percent of a sample to decay. In the next half-life, 50 percent of what remains will decay. Thus, we see that, with the horizontal axis representing time in half-lives, we end up with ½ → ¼ → ⅛ → 1/16 → 1/32 of the sample remaining after five half-lives.

half-life (another flip), half of the remaining ones will decay, and so forth. Thus, we can write that the number of nuclei still left at any time T compared to the number we started with at time $T = 0$, as

$$N(T) = N(T = 0) \times (\frac{1}{2})^{T/t_{\frac{1}{2}}}$$

where $t_{\frac{1}{2}}$ is the isotope's half-life. Using the case above, at $T = 4$ hours and $t_{\frac{1}{2}} = 1$ hour, $N(4 \text{ hours}) = 10,000 \times (\frac{1}{2})^{4/1} = 10,000/16 = 625$. By midnight $T/t_{\frac{1}{2}}$ would be 12/1 and $(\frac{1}{2})^{12} = 1/4,096$, so we would expect only 10,000/4,096 or about two or three nuclei left undecayed; by 3 a.m., there would probably be none of the original sample left at all.

Radioactive isotopes have an enormous range of half-lives, running from 0.000000000000000000000023 seconds (2.3×10^{-23} s or 23 yoctoseconds) for

Hydrogen with six neutrons (^{7}H) to the 2,200,000,000,000,000,000,000,000 years (2.2 × 10^{24} years or 2.2 yottayears—yes, yottayears is a lotta years) of Tellurium-128. In general, the lifetimes roughly correlate with how far from the line of stability the isotope lies; for example, Te-124, -125, and -126 are all snugly nestled in the valley of stability, and ^{128}Te is not far off, whereas only ^{1}H and ^{2}H are stable, so ^{7}H is *way* off the curve.

IMPERTURBABLE CLOCKS

The reason radioactive isotopes are so useful in discovering things about the past is that their decay rates are almost completely imperturbable. You can pour acid on a sample of Uranium, heat it to a million degrees, freeze it to near absolute zero, place it in a strong electric or magnetic field, run over it with a tank—whatever you like, and you will not change the half-life by one iota. Such reliable timepieces are hard to find anywhere else in nature or in our technology.

We use these radioactive clocks in several ways, and we will explore these in more detail in upcoming chapters. Briefly, if one knows the number of atoms present at the start time, one simply needs to count the number left at the moment of observation and, using the known half-life, apply the equation above to find *T*. For example, when a tree is alive, it takes in all the common isotopes of Carbon from the air and incorporates them into its cellulose molecules. When the tree is cut down, the ^{12}C and ^{13}C remain in the wood, but the ^{14}C content is decaying. If one finds a log used in an ancient dwelling and discovers only half the expected amount of ^{14}C is present, one knows the tree was felled 5,730 years ago. (See chapter 8 for details, including the small corrections to this dating technique that are required.)

If one doesn't know the number of atoms the object of interest started with, but one is using a simple decay mode in which one radioactive isotope decays to a single stable isotope, none of which was present in the original sample, one can just take the ratio of the two, and read the time off the graph, as in figure 6.5. This is called an accumulation clock. If unknown quantities of both the parent and the daughter nuclei are present in the beginning, one can take the ratio of both to another, stable isotope of the radioactive species and, using a curve called an isochrone, derive the age of the object (see chapter 15 for details). By choosing isotopes with appropriate half-lives—from hundreds to tens of thousands of

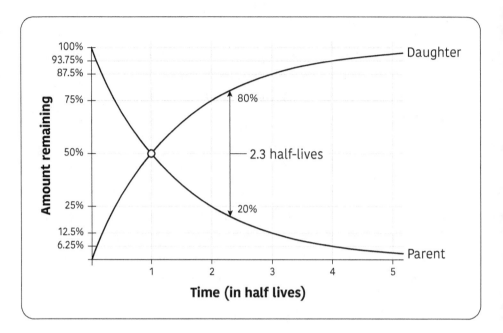

6.5 As the parent nucleus decays, the daughter nucleus builds up in direct proportion. For example, after one half-life, the sample will consist of 50 percent parent nuclei and 50 percent daughters. Assuming none of the daughter escapes (and there was none there to begin with), measuring the exact ratio of parent to daughter gives a unique point on the time axis for the age of the sample. In this case, the vertical line is at parent = 20 percent, daughter = 80 percent, so the sample is 2.30 half-lives old.

years for human artifacts, thousands to millions of years for climate studies, and up to billions of years for exploring the origin of the solar system and the universe, we have clocks available to tell time throughout the history of the cosmos.

I said "almost" completely imperturbable above, but you should feel free to ignore the "almost" in all the cases of interest we will study. For completeness, however, I'll cite the exceptions here. The most important exception is in the decay mode dubbed electron capture. Recall that this occurs when one of the orbiting electrons in an atom wanders too close to the nucleus and is captured, neutralizing one of the protons by turning it into a neutron. Because the atom must *have* electrons for one to wander too close, one can stabilize an electron-capture nucleus if one strips off all the electrons by ionizing the atom completely. A less dramatic step would be simply to alter the orbits of the electrons by surrounding the atom with other atoms or molecules. For example, the electron-capture half-life of Beryllium-7 was lengthened by 0.9 percent when the ^{7}Be atom was surrounded by Palladium atoms.[3] In general, however, we will be more than happy with an accuracy in our dating of 1 percent, so this is of no practical concern. Finally, the observed half-life of a nucleus can be changed if one changes the rate of time itself, either by accelerating a particle close to the speed of light or bringing it near to the event horizon of a black hole. According to Einstein's theory of relativity, time slows down in both cases, and that means the half-life of the nucleus will appear to us observers as being longer. The first effect has been demonstrated in particle accelerator experiments; the black hole experiment has yet to be undertaken. Neither circumstance, however, will be applicable to any of the history we wish to reconstruct.

With our understanding of subatomic particles, nuclei, atoms, and molecules now complete, we are ready to use these tiny tools in our project to reconstruct history. Let the stories begin.

CHAPTER 7

Stolen and Forged

Forensic Art History

Richly illustrated medieval paintings, choir books, and colorful miniatures from the fifteen century, clearly executed on 500-year-old wood, velum, and parchment, can currently be found in the collections of several of the world's leading museums. How can the colors be so bright after half a millennium? The nineteenth-century American painter Ralph Albert Blakelock produced hundreds of paintings but was so unsuccessful in selling them that he fell into a depression that led to his institutionalization. Thereafter, the prices of his paintings soared—but were they all his paintings? Nearly fifty works by the German surrealist Max Ernst, the expressionist Heinrich Campendonk, and the French painter Fernand Leger, along with other paintings from the first decades of the twentieth century, were sold at auction for a total of $36 million in the 1990s and early 2000s. Who was the mysterious collector that had amassed such a gallery of work? Then, of course, there are those Khmer guardians, statues that long ago were relieved of their feet. Mysteries all. And all readily solved by questioning the atoms involved.

THE FOOTLESS GUARDIANS

Footless they were, but when the Met first acquired the two kneeling attendants in the late 1980s, their heads and torsos were also separated. Collecting four separate pieces over five years, the museum acquired the available parts and put them together in 1993. The feet, however, remained missing.

Meanwhile Federico Caro of the museum's science department was studying the sandstone quarries around central Cambodia. With twenty samples from the quarries near Koh Ker and two dozen more from the Kulen Mountains half-way between Angkor and Koh Ker, he assessed the minute concentrations of twenty-three separate elements ranging from number 4, Beryllium, to number 92, Uranium. He showed how these trace elements were generally similar to those in other sandstones used in the Khmer cities and temples of 1,000 years ago. But the data also revealed that the concentrations of element number 21, Scandium, and element number 23, Vanadium, present in only ten and sixty parts per million, respectively, while tightly correlated across the various samples from his two sites, differed slightly from the ratios seen in the Angkor temples, suggesting different quarries were used in those buildings.[1]

One hundred kilometers from the famous temples of Angkor Wat, the Koh Ker area is of particular interest in Khmer history. King Jeyavarman IV moved the capital from Angkor to Koh Ker in 921 AD as part of a dispute with a late king's sons, who also claimed the throne; they died in 928 and the new king ruled unchallenged until his own death in 941. His elaborate new capital was completed in less than two decades, suggesting a substantial amount of capital and labor under Jeyavarman IV's control. The quarries Caro sampled provided the raw materials for the buildings and sculptures gracing the new city. A sample of the stone from the kneeling attendants could reveal the location from which they had come.

Before the difficult decision was taken to cut a sample from the statues for analysis, however, traditional archeology intervened. At the Western gate to the main temple in Koh Ker, the attendants' feet were found. Pursuant to an agreement signed in 2013, the museum returned the two statues so that they might resume their 1,100-year guard duty in Jeyavarman IV's capital.

THE SPANISH FORGER

In 1930, the Metropolitan Museum was considering a significant purchase. The medieval painting in question had been attributed by the English art historian Sir Lionel Cust, former director of the British National Portrait Gallery, to the fifteenth-century Castilian artist Jorge Ingles. The painting, entitled *The Betrothal of Saint Ursula*, shows a scene of a happy couple, surrounded by more than a dozen courtiers, standing in front of a castle, with several boats drifting on a lake in the background. Measuring 30″ × 24″, it is painted on wood and has the characteristic cracks to be expected of a 500-year-old painting; otherwise, it is in excellent condition. The asking price was 30,000 British pounds. Before completing the purchase, the Met's board of trustees asked Count Umberto Gnoli, another distinguished art historian who was also their purchasing agent, for a second opinion. He turned to Belle da Costa Greene, head librarian of the Pierpoint Morgan Library, two and a half miles south of the Met in Manhattan.

Greene declared the painting a forgery, and the museum declined the purchase. Inspired by the original attribution, she dubbed its creator the "Spanish Forger," although it is now thought most likely that the real forger worked in Paris at the end of the nineteenth century and the beginning of the twentieth. Over 350 works, many in the collections of the world's leading museums (including both the Met and the Morgan Library) are now attributed to this still unknown, but clearly prolific, forger.

Greene's analysis was based on art historical considerations, as was Cust's original authentication. Further such disagreements among experts are recounted below. Indeed, Greene had authenticated as genuine an illustrated medieval liturgy book that the Morgan Library had acquired twenty years earlier, although she subsequently came to recognize that artifact as the work of the same forger.

Is there any way our unbiased atomic historians could give us a definitive answer as to the provenance of these works of art? Indeed, there is. It's called autoradiography, or neutron activation autoradiography, to be precise. Neutrons, as you will recall, have no electric charge. Their flight through space is thus unaffected by any electrons, protons, or magnetic fields they may encounter. But they do find a comfortable home in the atomic nucleus, where they experience the powerful embrace of the strong nuclear force. Adding a neutron to an atomic nucleus does not, of course, change the atom's identity—^{12}C and ^{13}C are both still Carbon—but it does create a new isotope that may or may not be unstable to radioactive decay (e.g., ^{13}C + n = ^{14}C, which, as we saw in chapter 6, will undergo beta decay).

Beginning in the 1960s, the graphite nuclear reactor of Brookhaven National Laboratory on Long Island, New York, was used to examine paintings using neutron activation. The reactor generates a large number of neutrons (millions to trillions per square centimeter per second) that are moving with modest velocities. (With typical energies of 0.025 eV, they move at speeds comparable to the air molecules in the room and thus are dubbed "thermal neutrons.") A tiny fraction of these neutrons collides head-on with the nucleus of an atom in the painting and are captured by the strong force they feel, transforming the original atom into its heavier isotope.

For example, the only stable isotope of element number 11, Sodium (^{23}Na), makes up 100 percent of naturally occurring Sodium. If this nucleus were to capture a neutron, it would become the radioactive nucleus ^{24}Na, which is unstable to beta decay—it wants to get back closer to the valley of stability by emitting an electron from the nucleus and equalizing its number of protons and neutrons; this shifts it one step up in the Periodic Table to element number 12, Magnesium (^{24}Mg). The half-life of this decay is 15 hours. As in most such decays, this leaves the Magnesium nucleus in an excited state, which then undergoes a gamma decay to the ground state, emitting photons with energies of 2.75 and 1.37 MeV. Thus,

$$^{23}\text{Na} + \text{n} \rightarrow {}^{24}\text{Na} \rightarrow {}^{24}\text{Mg}^* + \text{e}^- + \bar{\nu}_e \rightarrow {}^{24}\text{Mg} + \gamma\,(2.75, 1.37 \text{ MeV})$$

Adding a single neutron to most stable isotopes produces similar chains of events. Note that, following this sequence, the net result is that (1) a high-energy electron flies off the painting, (2) gamma rays with very specific energies characteristic of the new element are emitted, and (3) the original atom is transformed to a new atom one step up in the Periodic Table.

It may seem a matter for concern that, through neutron illumination, one has changed the elements in a potentially precious work of art. But how many such atoms are changed?

In the Brookhaven project that analyzed seven different works attributed to the Spanish Forger, the reactor was used to irradiate the objects of interest placed 2 feet away. Each square centimeter of the artwork received about 1 billion (10^9) neutrons each second for roughly 90 minutes. For a six-inch page of an illuminated manuscript, that's a total of more than 1,000 trillion (10^{15}) neutrons. Recall, however, that unless a neutron comes within 10^{-14} cm of the nucleus, it doesn't feel the strong force's attraction at all and so passes right through the painting. Taking the tiny size of the targets into account, 99.9999999999 percent of the neutrons pass through and only five or six out of every trillion atoms in the painting are transformed by a neutron capture.

To picture how insignificant this is, imagine a warehouse covering an entire New York City block (260 feet × 900 feet). The building is twenty stories tall and filled completely with blue marbles—each representing one atom in the painting. The neutron irradiation changes five or six of those marbles to red. If 2,000 people worked eight hours a day, five days a week and managed to grab one marble each second and chuck it out of the window if it's blue, it would take them more than sixty years to find the five red marbles. And if the warehouse were next to Central Park, the discarded marbles would cover the entire park knee-deep. It thus seems fair to call autoradiography a nondestructive analysis technique.

Once radioactive nuclei have been created, they decay with their various individual half-lives. To show exactly where in each painting specific elements are located, a piece of film is applied to the front of the painting so that the exiting electrons expose the film. The film remains in place for various lengths of time, starting at different intervals after the time of irradiation, in order to detect the presence of different elements with different half-lives. In the case of the Spanish Forger's work, the elements of interest (with their respective half-lives) were Manganese (2.58 hours), Copper (12.7 hours), Sodium (15 hours), Arsenic (26.3 hours), Gold (2.7 days), Chromium (27.7 days), Mercury (46.7 days), Antimony (60.2 days), and Zinc (244 days).[2] The films were in place at the following intervals after exposure (for the following times): one day (one day), four days (two days), seven days (two days), and twenty-two days (five days).

One of the manuscripts, a book of hours from the 1400s acquired by the Morgan Library in 1900, before Greene's time, includes twelve calendar paintings and six illuminations, one of which is a scene with Mary Magdalene. The book itself is indeed 600 years old, and it was speculated that while the leaves and flowers gracing the borders of each page are original, the illuminations were added by the Spanish Forger. The audoradiographs revealed that the apparently indistinguishable blue paint used for flowers in the border (which contained Copper) was different from the blue used for the sky (which contained Sodium). While the half-lives of these two isotopes are similar, so we would expect equal intensities, Copper has a much greater ability to absorb neutrons and thus shows up more intensely. Blue pigment containing Copper was available in medieval times, but the blue sky was likely painted with ultramarine (with the chemical formula $Na_7Al_6Si_6O_{24}S_3$), a sodium-containing pigment first produced in the nineteenth century.

The film exposed between seven and nine days after irradiation shows a difference between the green leaves in the border and the green grass in the illustration, the latter being less intense. By the time of the exposure between 22 and 27 days, the green grass is completely inactive while the border leaves are still present. This is consistent with a pigment containing arsenic in the illustration; its half-life of only 26.3 hours would mean twenty half-lives have passed by 22 days and only one millionth of the induced activity remains.[3] Indeed, emerald green (copper-aceto-arsenite), contains arsenic. It was first made available to artists in 1818 and was considered the finest green pigment throughout the nineteenth century. It is poisonous, however, and was phased out of use in the early twentieth century.[4]

In addition to the variation in film exposure owing to the electrons shot out on schedules adhering to the various half-lives of the different induced radioactive isotopes, as noted above, the end products of the decays are left in an excited state and promptly emit gamma rays. Their energies correspond to the unique set of excited nuclear states for each element and thus provide an unambiguous signature for that element's presence. Because these photons are ejected in random directions, however, it is not possible to identify where in the painting they originate. Thus, the audioradiographic film and the spectrometer sorting the gamma rays by energy work in tandem: the spectrometer assures us that arsenic is present, and the film shows us where it is. In the case of the illustration of Mary Magdalene in the book of hours, the hypothesis that the borders were original while the illustration was forged proved to be correct.

THE BLAKELOCK OEUVRE

Forgeries of long-dead artists risk discovery through the changing composition of artists' pigments or the dating of the wood or canvas on which they are painted (see chapter 8). Faked works of contemporary artists, however, require subtler investigation, although some of the same nondestructive tools can be applied.

Ralph Albert Blakelock is reputed to be the most forged American artist. He began his career in New York in the early 1870s, after returning from an itinerant three years in the American West. He was largely self-taught, and while his early work is reminiscent of the Hudson River School, he evolved a unique, romantic style typified by dreamy, moonlit scenes. While highly productive as an artist, he was a feckless businessman, and his work fell far short of generating the means necessary to keep his wife and nine children fed and clothed. He had a mental breakdown in 1891 and by 1899 was permanently institutionalized. His departure from the scene saw a dramatic rise in the critical and financial success of his work, which began commanding four-, and then five-figure sums. With this high demand and no ongoing supply, forgeries proliferated, including some by his daughter Marian (who herself was institutionalized).

In the 1970s, Maurice Cotter and his colleagues began an extensive examination of Blakelock's work, along with that of his contemporaries, including some known forgeries.[5] A total of forty-five paintings were subject to autoradiography as well as X-rays. Taking an X-ray of a painting works the same way it does in the radiologist's office. A beam of X-ray photons passes through the painting and is detected on the other side, either on film or, more likely today, by an electronic detector. As with the image of a broken bone seen through skin and muscle, X-rays can reveal hidden layers in a painting; in addition, because each atom absorbs X-rays at the specific energies corresponding to its inner electron energy-level spacings, element identification is possible.

To assess the authenticity of Blakelock's oeuvre, the Cotter team began with a painting in the collection of the Smithsonian Institution, *Moonrise*, whose provenance was certain, traced back directly to the artist himself. An X-ray of this piece showed a technique characteristic of Blakelock's work in which he used a palette knife to apply uneven layers of Lead-white pigment as a base, in part to smooth the rough surface of the wood, and in part to act as an underpainting; in this work, what will become the brightest moonlight is underlain with the thickest layer of white paint. Because Lead is an efficient absorber of X-rays (recall that heavy apron the dentist puts in your lap when she X-rays your teeth), the thickest layers appear dark in the X-ray.

An autoradiograph taken a few hours after irradiation revealed a Manganese-based pigment laid down over the Lead-white base that preserves the lumpy appearance. Another radiograph taken several days later showed an Arsenic-containing pigment worked into the painting, perhaps with the wooden end of the brush, to highlight the light coming through the trees. Yet another image exposed between fourteen and twenty-four days after irradiation showed a thin layer of a Mercury-containing pigment brushed on as a glaze to achieve a reddish hue. Such final tonal layers are typical in Blakelock's mature work.

Another of the works these authors studied is a painting on canvas entitled *A Woman in Red*. The X-ray shows that, unlike in Blakelock's work, the priming layer is very thin, and thus the weave of the canvas is still easily seen. In a radiograph taken nine to nineteen days after irradiation, the woman's red gown is completely saturated. This is attributed to the decay of the radioactive isotope of Mercury ^{203}Hg, formed from the capture of a neutron by the most common stable isotope ^{202}Hg, which then undergoes a decay to ^{203}Tl; the pigment vermillion is mercuric sulfide (HgS). The application of the paint is quite different from that in the *Moonrise* painting. In the lower left of the painting, a partially scraped off signature shows up in the radiograph; comparison with a signed work by Marian Blakelock leaves little doubt that the daughter here replaced her own signature with that of her father.

Locked away in the Middletown State Homeopathic Hospital in Orange County, New York, Ralph Albert Blakelock was unaware of the meteoric rise of his reputation as an artist. In 1916, one of his landscape paintings fetched $20,000 at auction (more than half a million in today's dollars), a record at that time for a living American artist.[6] A reporter for the *New York Tribune* discovered Blakelock's whereabouts and, for the first time in seventeen years, brought him to Manhattan to see a gallery retrospective of his work. Thirty years later, the reporter acknowledged that Blakelock had told him at the time that a number of works in the show were forgeries but, being uncertain as to Blakelock's mental acuity—as well as being unwilling to ruin a feel-good story—he had omitted this fact in his account of the gallery visit.[7]

THE GERMAN FORGERIES AND PIXE

In November 2011, Wolfgang Beltracchi along with his wife, his sister-in-law, and her husband were convicted of forging fourteen European surrealist and expressionist paintings that sold in the putatively reputable art market for more than 16 million euros ($22 million at the 2011 exchange rate). Left out of the charges were at least another forty-four forgeries identified by German investigators; together, these elevate the case to the largest art fraud of the past seventy-five years.

The paintings were sold at auction in Germany in the early 1990s, and all were certified as genuine originals by art historians. Christie's auction house participated in some of the sales, lending credibility to the works. The comedian Steve Martin bought one of the works by the painter Heinrich Campendonk for $850,000.

Beltracchi's wife had concocted a clever provenance story—that her grandfather had purchased the works early in the twentieth century from the German Jewish art dealer Alfred Flechtheim, whose business was seized by the Nazis in the 1930s. Her grandfather, Werner Jaegers, was said to have hidden his collection in a mountain cave outside Cologne so they would not be stolen or destroyed by the Nazis (who considered the works of that period "degenerate"). In fact, Jaegers had never bought a single painting in his life, although the Beltracchis went so far as to have photographs taken in period clothing with the fake paintings hanging in the background.

Another Campendonk work, sold at the Lempertz auction house in Cologne in 2008 for $2.5 million, was the key to uncovering the fraud when the new owner had it tested with proton-induced X-ray emission (PIXE). PIXE is another completely nondestructive technique that takes advantage of the unique fingerprint every atom has in the set of specific wavelengths its electron transitions produce. PIXE uses an accelerator to create a beam of protons with energies (typically up to a few million eV) sufficient to ionize the atoms in a painting by ejecting one of the innermost electrons from its orbit. In response, the outer electrons cascade down to fill the electron hole, emitting a unique set of wavelengths characteristic of each element. While a similar process can be effected by illuminating the painting with sufficiently energetic X-ray photons (called X-ray fluorescence), PIXE has the advantage of being able to vary the beam energy continuously and thus select the depth of penetration the protons will achieve (allowing interrogation of the various layers of paint as well as the sizing on the canvas and the varnish overlain). In addition, one can focus the beam to illuminate specific microscopic spots on the work; beam sizes as small as 0.2 mm (a few times the width of a human hair) are not atypical.

PIXE examination of the Campendonk painting showed that it contained significant amounts of Titanium, used in the pigment titanium white that was first produced seven years after the work was supposedly created, classifying it as an unambiguous forgery. This technique leaves the painting completely unaffected—except, of course, for its price.

PIXE APPLIED

Three-dimensional artifacts are also the subject of forgers' interest. In China, jade items found in ancient tombs are highly valued. After thousands of years in contact with soil and water, white jade develops a luminescent patina known in China as "chicken bone white." To achieve this appearance quickly for forged modern pieces, the objects are soaked in either hydrochloric acid (HCl) or sulfuric acid (H_2SO_4). This leaves a telltale excess of either Chlorine or Sulfur that is easily picked up by PIXE scans. For example, while ancient jade artifacts have Chlorine concentrations of 0.01 to 0.036 percent, modern fakes were ten times higher, ranging from 0.1 percent to 0.36 percent.[8]

PIXE analysis can be exquisitely sensitive to subtle chemical changes; thus, it, has been applied to many areas of art and archeology. One of its more remarkable applications is its use in the attempt to place in chronological order Galileo's undated notes and drawings related to the development of his theory of motion. To assess the exact composition of the ink in dated letters and financial documents and to compare it to that on the undated pages of Galileo's notes on the physics of motion, an Italian team has compared the ratios of the elements Zinc/Iron, Copper/Iron, Zinc/Copper, and Iron/Lead, and managed to group documents by subtle changes in the contents of Galileo's inkwell, thus adding significantly to the scholarship on one of his most important contributions to modern science.[9]

With completely nondestructive techniques, we can dramatically enhance the field of art history by querying directly the atoms and molecules from which art is made, revealing provenance and perfidy, technique and time of creation. Applying such approaches to sacred texts and artifacts, as well as to architectural and artistic creations from both historic and prehistoric times, allows us to extend our atomic reconstruction of human endeavors to long before modern civilization emerged.

CHAPTER 8

The Carbon Clock

Pinning Down Dates

"**C**arbon dating suggests early Quran is older than Mohammad."[1] This headline appeared around the world in the summer of 2015, provoking consternation in the Muslim world, interest among historians, and irritation in me. Rather than an example of the type of skullduggery uncovered in the last chapter, this story is simply a matter of egregiously bad headline writing coupled with one of my pet peeves, innumeracy. The incident does offer the opportunity, however, to illustrate a very important dating technique that we will use frequently in the ensuing chapters and so represents the next step in our quest to turn atoms into accurate historians.

DATING PARCHMENT

In early 2015, a few leaves of a Quran, including portions of Suras (chapters) 18, 19, and 20, were discovered in the library at the University of Birmingham tucked inside another ancient copy of the Quran first acquired in the 1920s. The radiocarbon accelerator unit at the University of Oxford used a tiny fragment of the parchment from these folios and found that they date to the years between 568 and 645 AD. Because Mohammad lived between 570 and 632 AD, the clickbait-focused headline writer felt justified in writing "Carbon dating suggests early Quran is older than Mohammad," thus "suggesting" this version of the Islamic holy text could have preceded the prophet, who received it as a revelation. This is an antiscientific "suggestion" for reasons I will describe below. But first, let's consider how one can unambiguously date a 1,400-year-old document to within a few decades.

As described in chapter 6, ^{14}C is a radioactive isotope of Carbon that undergoes a beta decay to ^{14}N with a half-life of 5,730 years (the half-life's uncertainty is less than 1 percent). Because all living things on Earth are based on molecules involving Carbon, they are, while alive, constantly absorbing that atom from the environment (plants by sucking CO_2 out of the air, animals by eating plants—or other animals who eat plants). As we shall see in chapter 10, ^{14}C is heavier than the far more common ^{12}C, so plants discriminate against it in building their molecules, but including some of the heavier isotope is unavoidable.

As soon as a plant or animal dies, it stops incorporating Carbon from the environment. The two stable isotopes, ^{12}C and ^{13}C, remain in place, but the ^{14}C is always decaying and, without new replacements, slowly but surely decreases in abundance with respect to its stable siblings. After 5,730 years (one half-life), the ratio of ^{14}C/^{12}C will be one half the value it had when the substance was alive. As outlined in chapter 6, the number of atoms, N, left after a given amount of time, T, since death, $N(T)$, is given by the number of atoms around at the time of death, $N(T = 0)$, times the fraction (½) raised to the power $T/t_{½}$:

$$N(T) = N(T = 0) \times (½)^{T/t_{½}}$$

where $t_{½}$ is the isotope's half-life. After one half-life has elapsed $T = t_{½}$, so $T/t_{½}$ = 1, $(½)^1 = (½)$, and thus $N(t_{1/2}) = N(T = 0) \times (½)$—one half of the atoms are left. If one can measure the number of ^{14}C atoms in a sample and one knows how many there were at the start, one can invert the equation and solve for T, the object's age. In the case of the ancient Quran, 84.3 percent of the original number of ^{14}C atoms were present, leading to an age of 1,408 years before 2015, or roughly 607 AD.

At this point, you should have several questions. First, how can one possibly know how many ^{14}C atoms were present when the sheep, from which the parchment in question was produced, was alive? Given that the half-life of ^{14}C is only 5,730 years, how can there be any of that isotope left around at all if the Earth is 4.5 billion years old? And finally, how can one possibly measure the ^{14}C/^{12}C ratio from a tiny scrap of parchment? Given the importance of ^{14}C dating in much of what follows (not to mention radioactive dating more generally), it is worth taking some time to answer each of these questions.

WHENCE ^{14}C?

First, why does ^{14}C still exist in the environment? As noted in chapter 6, this isotope is rare; it is about one trillionth of the amount of ^{12}C in the world. But if we only had the original ^{14}C allotment from when the planet formed, there would certainly be none whatsoever left by now—the fraction (½) raised to the ratio of the age of the Earth to the half-life of the isotope would mean multiplying ½ × ½ 797,033 times, a number as close to zero as you are ever likely to encounter.

In fact, our supply of ^{14}C is continuously replenished from the bombardment of our atmosphere by extremely high-energy particles from outer space called cosmic rays, which were discovered in 1911 by the Austrian physicist Victor Hess in a series of daring balloon ascents to high altitude. Cosmic rays consist of electrons, protons, and heavier atomic nuclei accelerated in interstellar space to speeds very close to the speed of light (see chapter 16) so that, despite their miniscule masses, they carry enormous energies; the most energetic cosmic rays detected at Earth pack the wallop of a professional tennis player's serve in a single proton.

When a cosmic ray particle, having traveled thousands of years through the galaxy, smacks into an atomic nucleus in the Earth's upper atmosphere (roughly 30 km above the surface), it shatters the nucleus into its constituent particles, liberating lots of neutrons. These fast neutrons in turn collide with other atomic nuclei, transforming them into new isotopes and other elements. In particular, a fast neutron (n) encountering a Nitrogen atom (^{14}N, the atmosphere's dominant constituent) can knock out a proton and settle into the nucleus in its place:

$$n + {}^{14}N \rightarrow {}^{14}C + p$$

The nucleus still has fourteen particles but one fewer proton, so it moves from place number 7 to place number 6 in the Periodic Table and becomes ^{14}C. The roughly constant rate of ^{14}C production through this process, coupled with the isotope's natural decay rate, leads to the observed ratio of a trillion ^{12}C nuclei for each nucleus of ^{14}C in the air today.

I say "roughly constant," but that's not good enough for accurate dating. I can directly measure the production rate today, but how do I know what it was 1,400 years ago before cosmic rays, let alone radioactive isotopes, were even imagined? In fact, the production rate does vary in ways both predictable and unpredictable for three separate reasons. It is only because we have an independent way of determining this rate that ^{14}C dating is practical.

CHANGING PRODUCTION RATES

The first consideration is the rate at which cosmic rays from deep space arrive at Earth. This will fluctuate on very long timescales as the solar system circles the galaxy's center (once every 240 million years or so) and passes through different regions of space. If we happen to pass by the location of a recent stellar explosion, the principal site of cosmic ray production, the rain of high-energy particles, along with the creation of ^{14}C will increase. On the timescales of interest to us (thousands to tens of thousands of years), however, it is safe to assume the arrival rate in the vicinity of the Earth for interstellar cosmic rays is reasonably constant.

A second factor that determines the rate at which these rays affect the atmosphere is the magnitude of both the Earth's magnetic field and the Sun's activity. The cosmic rays are, by definition, charged particles (either plus or minus), and charged particles interact strongly with a magnetic field. Indeed, the Earth's magnetic field (that makes your compass needle swing around to point north) is strong enough to deflect some of the cosmic rays entirely and to funnel many of the others along its lines of force onto the North and South Poles, creating the aurora borealis and aurora australis (the northern and southern lights).

It turns out that the strength (and even the direction) of the Earth's magnetic field is not constant. Produced by turbulent currents in the molten part of the Earth's interior, the field waxes and wanes in strength and wanders around in direction: at the moment, the magnetic north pole is 395 km south of the geographic North Pole, having moved 570 km north and 810 km west in the past twenty years.[2]

More important for the cosmic ray penetration, however, is the strength of the field. Over the last two centuries, the overall strength has been decreasing by about 6 percent per century;[3] a weaker field allows more cosmic rays to reach the Earth's surface, boosting ^{14}C production slightly. If this trend were to continue, the field would disappear within the next few thousand years. Looking over longer times, the kind of fluctuation recorded recently is not unusual, and the current strength is similar to the average strength over the past 7,000 years. On still longer timescales (hundreds of thousands to tens of millions of years), the field has been shown to disappear and come back upside down, with the magnetic south pole residing near the geographic North Pole, and vice versa. This is irrelevant for ^{14}C production rates of interest here because, given its relatively short half-life, the isotope is not useful for dating anything much older than 50,000 years.

The Sun's activity level also plays a role in modulating the number of cosmic rays arriving at Earth. When the Sun is particularly active, its magnetic wind extends beyond the Earth's orbit, minimizing the number of cosmic rays that reach the atmosphere by deflecting them into space. When the Sun is relatively quiet, its magnetic influence shrinks, and cosmic rays come raining in. By studying other radioactive isotopes produced by cosmic rays such as Beryllium (^{10}Be, $t_{\frac{1}{2}}$ = 1.6 million years), we can chart changes in solar activity, which may also have implications for Earth's climate (see chapter 11).

There is a countervailing trend with solar activity, however, that must also be considered. When the Sun is active, violent flares on its surface fling protons and electrons at Earth in the form of solar cosmic rays. Because the rate of flaring rises and falls with an eleven-year cycle, and the intensity of that cycle varies significantly over the centuries, ^{14}C production is affected. This effect is relatively small, however, with the maximum solar contribution typically ranging from less than 1 percent to 5 percent of the interstellar cosmic ray production.

Finally, human activity in the last two centuries have left a mark on the ^{14}C concentration that will need to be accounted for by future historians and archeologists. Coal, oil, and natural gas are all derived from plant material buried approximately 100 million years ago or more. Burning these fossil fuels today releases their Carbon to the atmosphere (in the form of CO_2) and alters the abundance ratios of its isotopes. We will explore this effect further in chapter 11, but we note here that this very old fossil Carbon contains no ^{14}C; it has all decayed long ago. Thus, in the year 2200, historians might well conclude that artifacts from today's civilization are older than they actually are because of the anthropogenically depressed ratio of ^{14}C/^{12}C in our atmosphere today.

There is another competing effect that these future scientists will also need to take into account, however. Between 1950 and 1963, a large number of nuclear weapons were tested above ground. The huge quantity of neutrons produced in each explosion interacted with Nitrogen just as cosmic ray–induced neutrons do and produced several tons of ^{14}C. The atmospheric concentration of ^{14}C nearly doubled, peaking around 1965. As the Carbon slowly mixed into land and sea, that pulse has decayed away, so that by 2021, the remaining atmospheric excess approached zero and the $^{14}C/^{12}C$ ratio continued its downward trend because of the burning of fossil fuels (see chapter 11).[4]

If the foregoing were all we knew about the rate of ^{14}C creation—a changing magnetic shield, variable solar activity, and human-induced changes—we'd be stuck with having to accept large uncertainties in the ages of the objects we wish to date: the starting ratio of ^{14}C would be highly uncertain. Fortunately, we can call on other natural phenomena that allow us to calibrate the ^{14}C production rate over the past 55,000 years, greatly improving the accuracy of our dating procedure.

CARBON CALIBRATION

Like all other living things, trees utilize Carbon from the air in constructing their wood, which is about 50 percent Carbon by mass. In temperate zones, trees have marked annual growth rings, adding one layer of wood to the tree's girth each year. They thus provide a perfect calendar—the outermost layer is from the current growing season, the next one in is from last year, the third one from two years ago, and so on. The study of these tree rings is called dendrochronology. Because trees are what they eat (see chapter 10 for an extended riff on this theme), the Carbon isotope ratios in a given ring's wood provide an exact measure of the Carbon ratios in the air that year.

While many trees live a century or less, some species have much longer life spans. The giant sequoia of the U.S. West Coast can live more than 2,000 years, and the bristlecone pines of the high Rockies include living members more than 5,000 years old. Using a small hollow drill called a Swedish borer, one can withdraw a sample from each of those trees' 5,000 annual growth rings and measure directly the $^{14}C/^{12}C$ ratio in each. Using dead trees preserved in ancient structures or buried in muddy river bottoms or bogs and matching their ring patterns where they overlap with living trees, dendrochronologists have constructed a continuous record extending back 13,910 years[5] before the present (BP)[6] in which every single year is represented by a sample of wood that reveals the ratios of the Carbon isotopes. Similar calibration using coral reefs and layered ocean and lake sediments extends the utility of Carbon dating back 55,000 years, beyond which the number of ^{14}C atoms remaining in a sample usually becomes too small to count reliably.

Over the past two millennia, the corrections to the derived ^{14}C ages flip back and forth from positive to negative: too old today (meaning there is a deficit of ^{14}C being produced, so there is less in the tree ring, implying less in the air and we infer an older age), too young 116 BP (implying an excess of production so there is more there than we should expect and we infer, assuming less decay, a younger age), too old at 142 BP, too young at 180 BP, and so forth, up to 2,421 BP (471 BC), at which point the ^{14}C ages remain too young through to the end of the record at 55,000 BP. The discrepancy is 200 years at 2,657 BP, 400 years at 4,136 BP, and 600 years at 5,015 BP, although the uncertainty in the correction is only ± 15 years, implying we can determine the age of a 5,000-year-old object to 0.3 percent. At the end of the tree-ring record 13,910 BP, the ^{14}C age is 1848 ± 28 years too young. The excess of ^{14}C (implying younger age estimates) continues to grow until about 25,000 years ago, when the correction is about 4,300 years. The peak excess occurs around 39,000 BP (5,000 years off) and then falls to half that discrepancy at 50,000 years. The uncertainty at this age is about ± 350 years, still considerably better than 1 percent accuracy.

These high-resolution, precise data have allowed the discovery of subtle effects in the atmospheric concentration of ^{14}C. For example, there is a small but well-established difference between the Northern and Southern Hemispheres. Small changes have also been seen in locations where deepwater upwelling in the ocean brings water (and dissolved CO_2) to the surface. Having spent thousands of years disconnected from exchanging gas with the atmosphere, this water has a low ^{14}C ratio because of fraction that has decayed and, when released to the air, lowers the ^{14}C ratio in the environment.

There are even ^{14}C anomalies identified with a single year, such as the Miyake event of 774–775 AD, which saw a sudden 1.2 percent spike in ^{14}C concentration (more than twenty times the typical year-to-year variation).[7] The onset of that event is not resolved but is almost certainly less than a year; after this sudden rise, the ^{14}C concentration slowly decayed back to the normal level over approximately twenty-five years. As further evidence that some event led to many high-energy particles striking the atmosphere in 774 AD, an Antarctic ice core shows a coincident spike in the radioactive isotope Beryllium-10, which is also produced by cosmic ray interactions with the atmosphere. Curiously, the Anglo-Saxon Chronicle for this year states,

This year the Northumbrians banished their king, Alred, from York at Easter-tide; and chose Ethelred, the son of Mull, for their lord, who reigned four winters. This year also appeared in the heavens a red crucifix, after sunset.[8]

The most likely explanation for this spike in ^{14}C production is a giant solar flare, perhaps 100 times more powerful than the largest flare observed in modern times (the Carrington event of 1859—see chapter 11 for the devastating consequences such an event would have today.) The "red crucifix" could be explained by a dramatic aurora borealis display (along with a healthy dose of medieval imagination).

The tree-ring calibration renders the difficult problem of knowing the ^{14}C/^{12}C ratio at the time of a plant or animal's death tractable, allowing a correction to ^{14}C dating accurate to roughly ten or twenty years. The derived ages thus have errors of a few tenths of a percent or less throughout the entirety of human civilization.

MEASUREMENTS ON TINY SAMPLES

Returning to the actual measurement of our ancient Quran, the final issue with dating is how to count the number of ^{14}C atoms present. When Carbon dating was invented in 1946 by Willard Libby, the only way to measure the number of ^{14}C atoms was to count the high-speed electron given off as each one of them underwent a beta decay (the electrons have a wide range of energies centered around 50,000 eV). Given the half-life of 5,730 years (or about 3 billion minutes), one needed 6 billion ^{14}C nuclei to expect even one decay per minute (because half of them decay in one half-life). The average ratio of ^{14}C/^{12}C is 1 in a trillion to start with, and one would need to measure 10,000 decays in order to reduce statistical errors to the 1 percent level.[9] Thus, this technique would require a sample of at least 15 grams of parchment—the mass of several pages of the manuscript.[10] No museum director wants to know the age of a precious object in her collection so badly that she is willing to allow it to be taken away to a physicist's laboratory for who knows what nefarious purposes.

Modern ^{14}C dating requires vastly smaller samples and counts the atoms directly, without waiting for them to decay, by employing an accelerator mass spectrometer (AMS; recall Francis Aston's invention of this device in chapter 5).

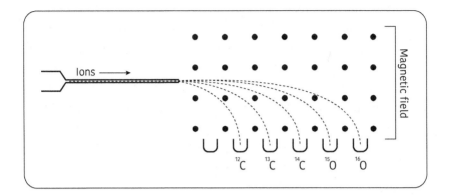

8.1 The operating principle of an accelerator mass spectrometer (AMS). The sample is vaporized and its atoms ionized so they are all positively charged ions. They are shot to the right, where they enter a magnetic field (perpendicular to the page, represented by small circles). Charged particles are deflected by a magnetic field, and the amount of deflection depends on the ion's mass. Thus, collection buckets can be placed at different distances from the ionizing source, and ions of different masses will be sorted, one mass for each bucket.

One extracts the Carbon by turning it to graphite (through burning or chemical techniques), then ionizing the atoms and accelerating them to high energies. This obviously requires the destruction of the sample, but the samples are thousands of times smaller, measured in milligrams rather than grams.

The high-energy beam of atoms is passed through a magnetic field that sorts the atoms by mass, with the more massive isotopes deflecting less (see figure 8.1). Samples as small as 0.01 grams are sufficient to obtain millions of ^{14}C atoms in a sample that is 35,000 years old.[11] As we shall see in subsequent chapters, AMS is an invaluable tool for counting both stable and radioactive isotopes used to reconstruct history.

THE AGE OF THE QURAN

Having accounted for the production of ^{14}C atoms, developed a calibration for their formation rate, and described the means to count them, we can now return to the misleading headline that began this chapter. Recall that the ^{14}C date was given as being between 568 and 645 AD. First, we should establish what that range means: it is a statement that the true age has a 95 percent chance of lying in that window and a 5 percent chance of being either older or younger. If you wanted greater than 99 percent confidence (the minimum we physicists would accept), you would have to expand the range to 549 to 663 AD[12] (recall that Mohammad died in 632). Furthermore, the date marks the time when the sheep or goat or calf whose skin was used to make the parchment died, not when the manuscript was written.

In fact, we have no technique for dating the ink, although by shooting X-rays at the manuscript to excite the electrons in the ink molecules and watching them jump back into place (emitting photons of wavelengths unique to each element the ink contains), it has been determined that the red ink was red-lead (an ink in use at least since 300 BC), and the brown was Iron gall made by mixing iron sulfate with the extract of oak galls (tree growths induced by wasp larvae), which has been used since the early fifth century. Nothing unusual there. By shining light of different wavelengths (both those we can see and those of ultraviolet and infrared light that lie beyond the range of human vision), scientists showed that, unlike many old manuscripts, there is no evidence for any previous writing on this parchment—it looks like an original skin prepared solely for this purpose.

While the initial version of the Quran was passed through oral tradition, it is known that the Caliph Abu Bakr had a written copy made between 632 and 634 AD immediately following Mohammad's death.[13] A subsequent Caliph, Uthman, is said to have commissioned the definitive edition during his reign, between 644 and 656 AD. Both 634 and 644 are within the window of acceptable ^{14}C ages, and the parchment could well have been prepared a few years before it was used, so there is absolutely no conflict with the date of the manuscript and the attribution of its words to Mohammad. The analysis does suggest these pages are from one of the earliest Qurans ever written, making them highly valuable, but it offers no reason to doubt the holy book's authorship.

Over the past seventy-five years, ^{14}C dating has been applied to dozens of substances to answer thousands of questions covering fields ranging from art history and archeology (see chapter 9), to human diet and agriculture (see chapter 10), the history of our climate (see chapter 11), the state of the oceans, and many more. Anything that was once alive (and even in one case, not alive—see chapter 9) is fair game: bone, charcoal, parchment, leather, linen and other natural textiles, seashells, beeswax, wine residues, pollen grains, and so on.[14] But while we are on the subject of religious artifacts, I will conclude this chapter with one of the more famous and controversial artifacts to which Carbon dating has been applied, the Shroud of Turin.

THE SHROUD OF TURIN: HISTORICAL HOAX

Celebrated for over half a millennium as the burial shroud of Christ, this piece of linen, 14 feet × 3.5 feet and woven in a herringbone pattern, has been housed in Turin Cathedral since 1578, nearly half a century after it was damaged in a chapel fire at Chambery in Savoy, the ancient dukedom that lies at the intersection of France, Italy, and Switzerland. The cloth bears the faint, negative image of a man with a beard and long straight hair, curiously consistent with medieval depictions of Jesus. There are faded red "bloodstains" at the hands and feet.[15]

For the past 125 years (beginning with the first photograph of the Shroud of Turin taken in 1898 that showed the image more clearly as a negative), countless scientific and not so scientific investigations have been conducted. The most definitive result from my perspective, however, occurred in 1988. Small samples of the linen were sent to three leading radiocarbon dating labs at ETH in Zurich, the University of Oxford, and the University of Arizona. The independently dated samples were all consistent with a flax harvest that occurred between 1260 and 1390 AD.

In turns out that these dates simply confirm what was already clear from the written historical record. The first mention of the shroud is in the 1350s in Lirey, France, where, we are told, it was proving to be a very lucrative relic for the local church dean, who charged pilgrims from all over Christendom to see it. The presiding bishop, Henri de Poitiers, was sufficiently irritated by this display (whether by jealousy or purer motives is not recorded) that he launched an investigation of the putative burial shroud and called a halt to the exhibitions.

Three decades later, it fell to his successor, Pierre d'Arcis, to interdict the reappearance of the shroud and send a letter to the Avignon pope that read, in part:

> The case, Holy Father, stands thus. Some time since . . . the Dean of a certain collegiate church, to wit, that of Lirey, falsely and deceitfully, being consumed with the passion of avarice, and not from any motive of devotion but only of gain, procured for his church a certain cloth, cunningly painted . . . the truth

being attested by the artist who had painted it, to wit, that it was a work of human skill and not miraculously wrought or bestowed.[16]

One might have thought the fact that the ^{14}C date is wholly consistent with the period during which the shroud first appeared, coupled with this written documentation, would be sufficient to put this centuries-old fraud to rest. But the controversy over the shroud rages on. No matter that shrouds of the first century Middle East do not have a herringbone weave, no matter that a molecular analysis of the "blood" showed it to be iron oxide (a pigment used by artists in the Middle Ages), and no matter that the image's right arm extends to an anatomically impossible length to discretely cover the figure's genitals, an endless number of studies continue to offer "verification" of the authenticity of the shroud.[17]

But the atoms do not lie, and they are unaffected by human sentiments. So I'm sticking with them as we continue our reconstruction of human history, as well as the history of the universe before it was troubled by humanity at all.

CHAPTER 9

History Without Words

Lime and Lead and Poop

C onfirming the age of dated documents through their atomic constituents is satisfying; the consistency provides evidence that the techniques we are using offer valid and valuable information. The cases for which no documentary evidence exists, however—whether as a consequence of lost records or from the works of preliterate peoples—is when our use of atomic historians is most illuminating. In this chapter, we will solve historical mysteries extending from the late 1600s back to some of the earliest cultural artifacts humans produced almost 50,000 years ago, as well as hominid tools more than ten times older than that.

DATING MORTAR

In chapter 8, we saw how ^{14}C can be used to date just about anything that was once alive and is well preserved. Here, we begin with a singular example of how Carbon dating has been applied to an inert building material, namely, mortar. The insights derived from this technique include the death of a Viking myth and the dating of medieval churches in Finland, as well as historically situating major architectural works from the Roman Empire both at its peak and following its demise.

Mortar, a crucial structural component of stone and brick buildings, was first used in ancient Mesopotamia and Egypt more than 4,000 years ago. Strictly speaking, it is incorrect to say that mortar was never alive. It is made from limestone, which itself is composed of the compressed skeletons of tiny ocean creatures called foraminifera (which we shall encounter again in chapter 11), corals, and mollusks with shells. As these creatures die and fall to the ocean floor, thick layers accumulate and, in response to the motions of the Earth's tectonic plates, are compressed into rocks and thrust up on land. The primary chemical constituent of these shells (and thus of limestone itself) is calcium carbonate, a molecule with one atom of Calcium, one of Carbon, and three of Oxygen: $CaCO_3$. In order to make mortar from this raw material, the following steps are required:

1. Heating the limestone to a temperature of at least 900°C to make quicklime.
2. Adding water to make a semiliquid suspension of the lime and possibly adding some crushed rock or sand.
3. Layering the wet mortar between the bricks or stones of the wall, and then allowing it to cure to hardness (over days to weeks).

What's happening chemically in these steps is as follows:

1. Heating the limestone drives off carbon dioxide (CO_2), turning $CaCO_3$ to CaO (quicklime).
2. Then adding H_2O to CaO leads to CaO_2H_2.
3. The mortar cures by releasing water (H_2O) to the atmosphere (drying) while sucking in CO_2 from the air to turn itself back into $CaCO_3$ (see figure 9.1).

The final step is the key to Carbon dating—the mortar is "breathing in" CO_2, just as plants do. In a matter of days to weeks, it has completed this process and so is effectively "dead" (no more breathing). All one must do to ascertain the time at which the mortar was applied is to count the ^{14}C and ^{12}C atoms and take their ratio, just as in normal Carbon dating.

The potential of this dating technique was recognized in the 1960s, but the initial attempts to apply it were largely unsuccessful, primarily because of impurities mixed into the mortar that could drastically bias the results. For example, unburned bits of limestone remaining in the mortar have a $^{14}C/^{12}C$ ratio of zero. Because all the ^{14}C has decayed away in the millions of years during which the limestone was forming, such material yielded ages drastically too old: the impurities reduced the $^{14}C/^{12}C$ ratio significantly. Sand or other grit mixed into the mortar prior to its application could contain bits of limestone or other Carbon-containing material that would also result in a misleading ratio.[1]

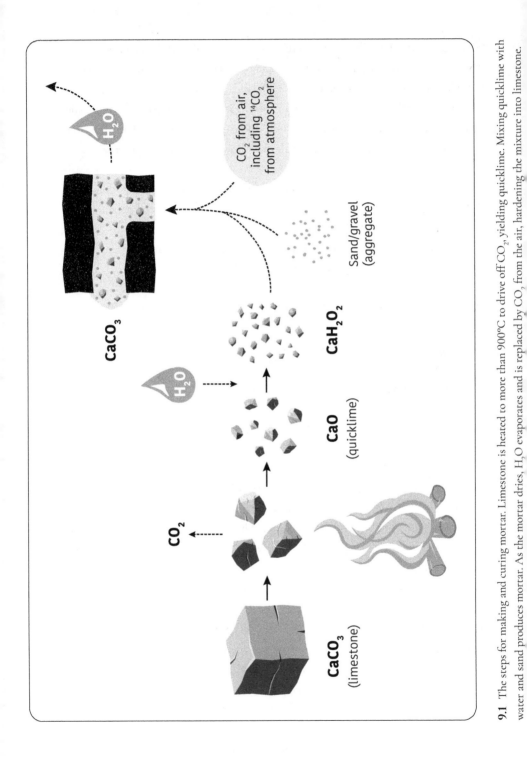

9.1 The steps for making and curing mortar. Limestone is heated to more than 900°C to drive off CO_2, yielding quicklime. Mixing quicklime with water and sand produces mortar. As the mortar dries, H_2O evaporates and is replaced by CO_2 from the air, hardening the mixture into limestone.

THE ALAND CHURCHES

In the 1980s, archeologist Kenneth Gustavsson and physicist Hogne Jungner set out to date the stone churches and Franciscan monastery of the Åland Islands, an archipelago off the coast of Finland. No written accounts of the construction of these churches existed, although they were assumed to have been built around 1450 AD. The radiocarbon laboratory of the University of Helsinki dated samples of mortar collected from the island of Kökar and returned the astonishing date of 1280 AD. Subsequent excavations around the church discovered jewelry that was roughly consistent with this earlier date.

Extraordinary claims require extraordinary evidence, and given the problems cited above with mortar dating, a ^{14}C date 170 years earlier than suggested by standard archeological wisdom called for an independent age estimate. Another dating technique that relies on the same cosmic rays that produce ^{14}C provided further evidence of the older date of construction: thermoluminescence. Thermoluminescence is just what it sounds like: light ("lumin") from heat ("thermo"). When a crystalline substance—either a natural crystal or a humanmade one like ceramic—is formed, its nominally perfect, regular structure inevitably contains defects, atoms or molecules not in exactly the right place and/or impurities that interrupt the regular crystal structure. These defects can produce small electric traps in which free electrons wandering through the material can get stuck.

Normally, electrons are bound to their respective atoms. But when a cosmic ray ploughs into matter, it can ionize lots of atoms and allow several freed electrons to escape into the material. In addition, natural radioactive isotopes within the material that undergo decay by spitting out high-speed electrons have the same effect. Thus, as time goes on and the exposure increases, more and more electrons find themselves in traps. If enough energy is added to free them from these traps, they can recombine with the atoms that are missing electrons and emit light.

In order to ascertain the age of an object (in the current case, a ceramic roof tile from the Franciscan monastery), we simply heat the material to a high temperature (approximately 500°C), making the atoms vibrate violently enough to liberate the electrons from their traps. We then record the amount of light emitted by these electrons as they rejoin atoms; the light intensity is proportional to the total radiation exposure, which in turn is simply proportional to the age of the object.

Because the number of trapping sites and the depth of the traps are unique to each substance, it is necessary to calibrate the sample by exposing it to a known amount of radiation for a known time interval and then repeating the thermo-luminescence experiment. When this technique was applied to the roof tiles on Kökar Island, the age suggested manufacture in the thirteenth century, consistent with the mortar's ^{14}C date of 1280.

With this success, a larger project to date all eight of the big churches in the Åland Islands was launched. In one, at Jomala, a timber in the bell tower was identified through tree ring matching to have been cut in 1281 AD, while the mortar from the walls of the church were dated to 1279–1290, a remarkable match that unambiguously determined the date of construction. Indeed, it appears all eight churches were built in the two decades between 1280 and 1300. The historical speculation is that the islands, situated halfway between modern Stockholm in Sweden and Turku in Finland (80 miles from each), enjoyed an economic burst of activity at that time by supplying the timber and lime mortar for the construction of those two cities.

Several refinements of the techniques for handling the mortar were required to obtain such precise dates: finer sieving to screen out impurities; electron bombardment of the sample to identify any hidden extraneous material through the characteristic wavelengths of light given off; and careful sampling of the outermost layers of the dried mortar, where contamination was typically the lowest. The high reliability of the dates derived allowed a multidisciplinary team of J. Hale (an archeologist), J. Heinemeier (a physicist), L. Lancaster (an archeologist), A. Lindroos (a geologist), and A. Ringbom (an art historian) to construct a detailed history of the Åland churches, including the various additions and renovations conducted during their 750-year history.

Working with colleagues, this team was also able to debunk the myth that a stone tower in Newport, Rhode Island, was a Viking structure from the eleventh century (it was a windmill base from 1680) and to resolve the date of construction of the Roman amphitheater in Merida, Spain (it was built 100 years after the 8 BC inscription in its wall and thus followed, rather than preceded, the construction of the Coliseum in Rome). They also reconstructed the history of a villa in Torre de Palma, 70 km west of Merida in Portugal. The chapel and its altar were built around 340 AD, while the Roman Empire was ruled by Constantine II, son of the emperor who adopted Christianity. But the structure was substantially enlarged in 580 AD long after the fall of the empire, when the Visigoths were in charge, demonstrating that much of the technology the empire had created lived on long after its demise.

ACCUMULATING LEAD

Even from periods in which plentiful historical records exist, our knowledge of the past is often still woefully incomplete. Absent hard information, speculation abounds. One such example is the oft-repeated assertion that Lead in the drinking water of ancient Rome precipitated the empire's decline. While ancient Romans clearly knew of and used Lead extensively in their building projects and utensils, they obviously had no way of discerning how much of it they ingested nor its harmful effect on the human body. A few years ago, H. Delile and his colleagues conducted a definitive study that shed light on this speculation, as well as revealing a history of Rome during and after the fall of the empire.[2]

There are four stable isotopes of Lead (Pb). ^{204}Pb is primordial, left over from the formation of the Earth, and accounts for roughly 1.4 percent of natural Lead deposits. The other three isotopes mark the end points of radioactive decay chains of the elements Uranium (^{238}U → ^{206}Pb and ^{235}U → ^{207}Pb) and Thorium (^{232}Th → ^{208}Pb); these three isotopes contribute roughly 24 percent, 22 percent, and 52 percent, respectively, to Lead ore, although the precise ratios depend on the exact concentration of their parent nuclei and thus serve as markers of the geological locations from whence they came.

Lead has been used for a variety of purposes for over 9,000 years; its low melting temperature of only 328°C makes it the easiest of the metals to smelt from its ore (called galena or lead sulfide—PbS). Originally used to make small sculptures and jewelry, it found use in a range of cultures for everything from a stimulant to cosmetics, a contraceptive, a sweetener, and coffin construction. In the Roman period, its highly ductile nature and resistance to corrosion made it a natural choice for the extensive network of aqueducts and pipes that supplied the capital (and other cities of the empire) with water. The Latin word for Lead is *plumbum*,[3] hence our word for plumbing, and the element's chemical symbol Pb. Indeed, at the height of the empire, the production rate of smelted Lead is estimated to have been 80,000 tons annually from mines in England, Spain, the Balkans, and Asia Minor. Over the empire's lifetime, nearly 15 million tons were used, an amount so great that it left a signature in the Greenland ice cap (see chapter 11) where an estimated 400 tons of airborne Lead pollution was deposited between 300 BC and 500 AD.

While Pliny the Elder (among others) writes of Lead fumes being toxic, there is little documentary evidence of Lead poisoning in the ancient world, and it is highly doubtful it occurred to anyone that the Lead pipes and cisterns (not to mention cooking utensils and wine storage containers) caused any harm. And, of course, there was no way to assess the Lead content of water or wine. This is where atomic history steps in.

Delile and his colleagues have made a direct measurement of the Lead content of the Roman water supply over a period of 1,000 years by drilling and extracting two 13-meter-long cores from the sediment in the harbor of imperial Rome at Portus and from the canal connecting the harbor to the Tiber River. In addition, the researchers obtained samples of Lead from the modern Tiber riverbed and from five Roman-era pipes. A total of ninety-five samples from different depths in the core and from various riverbed locations and pipes were assessed to determine their Lead isotope ratios. The core depths were dated by using ^{14}C dating of the organic material present throughout the cores.

By plotting the ratios of ^{204}Pb, ^{207}Pb, and ^{208}Pb to ^{206}Pb, a striking pattern emerges. Prior to the construction of the harbor at Portus around 100 AD, the ratios are constant, with values similar to modern-day values from the area—this is the naturally occurring Lead in the Tiber's waters characterized as arising from two sources: erosion of volcanic rocks from the Alban Hills just south of Rome and from limestones in the headwaters of the river in the Apennines to the east. Just after the harbor's creation in 112 AD, however, all three isotopic ratios shift to higher values, indicating the deposition of water laden with Lead from the pipes (which have the highest measured isotope ratios of the entire data set). The Lead deposits reach a peak in 200–250 AD and then fall to a middling value for 250 years until the early sixth century, when they rise again through 800 AD before falling back to the original background levels. These clear shifts correspond to specific events in the history of the Roman Empire: the harbor's creation around 100 AD; the decline of the population in the third and fourth centuries; and the repair of the water system by Flavius Belisarius, the Byzantine general who retook Rome from the Goths in 538 AD. The Arab sack of Rome in 846 AD saw the final destruction of the water system and the return of the Lead isotope ratios in the river to their natural background values.

From a quantitative analysis of the Lead content, the authors estimate that 3 percent of the total flow of the Tiber ran through the Romans' Lead pipes. This raised the Lead concentration by fortyfold in the imperial period. While this level of Lead pollution may sound frightening, it is less than the amount we live with today. From the few bone samples available (note that approximately 90 percent of ingested Lead ends up residing in the bones, so they are a good proxy for total exposure), it appears that the residents of Rome and other cities throughout the empire ingested Lead at about 45 percent the level we find in modern Europeans. It would appear, then, that we must look to causes other than Lead poisoning to explain the fall of Rome.

THE FIRST ART

The earliest human artistic creations—from a time long before writing existed—are found on the walls of caves and other rock surfaces. In 1940, several boys following their dog down a hole discovered the Lascaux Cave's "art gallery" featuring hundreds of colorful paintings of animals from prehistoric Europe. One of the first major applications of Libby's ^{14}C dating technique to a work of unknown age was conducted in 1951 on a sample of charcoal from the cave; it yielded an age of 15,500 years before the present (BP; see chapter 8). Subsequent dating of two other samples found ages of 16,000 and 17,200 years BP, respectively. More recently, accelerator mass spectrometer (AMS) dating of a fragment of a reindeer antler found in the cave gave an age between 18,600 and 18,900 years BP. Note that all of these are of secondary objects, not samples from the drawings themselves, and the span of more than 3,000 years is disturbingly imprecise.

As discussed in the introduction to this book, our atomic historians lack the cultural biases of humans, but they do have behaviors that can confound our use of them to reconstruct a quantitative history. In 1990, J. Russ and collaborators introduced a technique that cleverly avoids one of the interfering features we found above with mortar—the contamination of a sample with ancient inorganic Carbon. (Because much of the world's cave art is found in limestone caves, the excision of even milligram samples of paint can contain contaminating Carbon from the underlying rock.) Russ and colleagues applied their new technique to a rock painting from the Big Bend region of the Rio Grande River in southwest Texas.[4]

Most of the pigments used in rock painting (e.g., iron and manganese oxides) are inorganic and so contain no useful Carbon. But these pigments don't stick to a wall, so they must be mixed with materials that will both dissolve the pigments and stick to the surface. Chemical analysis of cave art from around the world has found that blood, urine, milk, honey, eggs, seed oils, and plant resins have all been used as media. All these substances are, of course, organic and thus can be used in ^{14}C dating. The trick is to extract only the organic Carbon without allowing any inorganic Carbon from the rock surface along for the ride.

The researchers used a vacuum chamber that is first thoroughly evacuated and then filled with pure Oxygen gas (O_2). By placing the sample in the chamber and heating it to 100°C, all of the organic (Carbon-containing) compounds are oxidized to water (H_2O) and Carbon Dioxide (CO_2). The relatively weak chemical bonds of the organic molecules succumb to collisions with the highly reactive Oxygen atoms at this temperature, but none of the Carbon from the limestone ($CaCO_3$) is disturbed (recall that separating the CO_2 from limestone requires heating it to at least 900°C). This direct sampling of the artist's medium avoids the ambiguity of dating charcoal or bones or other secondary bits of evidence. The water and CO_2 are then extracted from the chamber by a cold finger (water freezes at 0°C, and CO_2 freezes at −78°C). A total of 12 mg of pure Carbon was extracted from the residual CO_2 and dated with AMS. The result was an age of 3,865 ± 100 years BP, consistent with the broad range of ages expected from archeological evidence of this period.

OLDER AND OLDEST

For many decades, it was thought that the oldest scenes painted on cave walls were in Europe. Following the discovery of the caves at Lascaux, numerous other sites were found in Spain, France, and other European countries. In 1994, a huge cave was found in Chauvet-Pont-d'Arc, 40 miles north of Avignon in southern France. Carbon dating has suggested that the artwork was done during two different periods, the first between 33,000 and 37,000 years ago, and the second from 28,000 to 31,000 years ago.[5]

Curiously, Carbon dating of cave bear bones from over 200 skeletons found in the cavern overlap substantially with the first set of dates. These huge creatures were larger than modern grizzly bears, standing up to 12 feet tall and weighing over 2,000 pounds, and would be unlikely companions for human artists—perhaps they had a time-share arrangement.

These dates have received independent confirmation from thermoluminescence studies. In several places in the cave, the naturally beige limestone has been turned red by fire (whether accidentally or deliberately is unknown). As noted above, heating a material releases the trapped electrons, resetting the clock to zero radiation damage. Thermoluminescence dates were obtained for two samples from different parts of the cave complex, yielding values of 36,900 and 34,300 years ago with uncertainties of 2,000 to 3,000 years but within the first period of Carbon dates derived above.[6] The oldest of these dates follows by only a few millennia the arrival of modern *homo sapiens* in Europe.

Recently, claims have been made that art began in Europe more than twice as long ago, perpetrated not by *homo sapiens* but by Neanderthals. Red markings in a Spanish cave (a far cry from the elegant animal portrayals in the Chauvet and Lascaux paintings) have been dated to 65,000 years BP. However, several authors have questioned both the artifacts and the dates.[7]

Recent discoveries on the island of Sulawesi in Indonesia have eclipsed the oldest verified European paintings by a considerable margin. While none of the pigments have been dated directly, a very convincing isotopic dating technique provides a lower limit to the age of these paintings by examining natural mineral deposits that have accumulated on top of the artwork after it was completed.[8]

Speleothems (from the Greek, *spelaion* ["cave"] and *thema* ["deposit"]) is the collective name for the stalagmites, stalactites, and flowstone deposits that slowly accumulate as water drips into an existing cave, carrying with it dissolved minerals from its trip through the overlying soil and rock. Like the cave itself, speleothems are primarily composed of $CaCO_3$, but they include many other trace elements. In particular, the dripping water regularly includes easily soluble Uranium but excludes Thorium.

As noted earlier, all Uranium isotopes are radioactive, and the end point of their decay chains is Lead. But along the way, they produce Thorium (^{230}Th), which has a sufficiently long half-life (75,500 years) that we should expect some of this intermediate decay product to be present in the speleothem material. The decay chain looks like this:

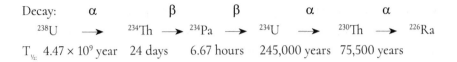

Decay: α   β   β   α   α

^{238}U $\longrightarrow$ ^{234}Th $\longrightarrow$ ^{234}Pa $\longrightarrow$ ^{234}U $\longrightarrow$ ^{230}Th $\longrightarrow$ ^{226}Ra

$T_{\frac{1}{2}}$: 4.47×10^9 year   24 days   6.67 hours   245,000 years   75,500 years

Given their short half-lives, ^{234}Th and ^{234}Pa are effectively nonexistent in a speleothem sample, but the ratios of ^{234}U and ^{230}Th to ^{238}U provide direct information on when the flowstone formed.

As in all such measurements (as we have learned by now), it is important to be careful about biases. For example, if there is some ^{230}Th in the rock layer, deposited by windblown dust impinging on the solidifying speleothem, the age we derive will be too old because an excess of ^{230}Th implies the ^{238}U has had longer to decay. The antidote here is to measure the ratio of ^{230}Th to ^{232}Th; ^{232}Th has a half-life three times the age of the Earth, so primordial Thorium naturally still exists in the planet's crust. Because of the constant supply of ^{230}Th from Uranium decay, however, the ratio of ^{230}Th to ^{232}Th on the surface is about 0.835. Thus, a speleothem with a large ratio of ^{230}Th to ^{232}Th must have most of its Thorium from Uranium decay—if the ratio is greater than 20, at most 0.825/20 or 4 percent was from contamination. In the samples derived from the Sulawesi speleothem, the ratios were even larger, ranging from 29 to 369, indicating negligible contamination.

Another concern arises from the possibility that ^{238}U and/or ^{234}U could have leached out of the speleothem after its formation. This would also lead to an age too old because of an underestimate of the amount of Uranium in the sample. Because the outer layer of the rock would be most susceptible to such leaching, we would expect the outermost layer to register as oldest when in fact it must be the youngest given the progressive layering of the dripping mineralized water. The authors checked this by taking five slices of the speleothem from the outside in (the innermost layer being closest to the pigments) and found in all cases that the ages increased going *in*ward, as would be expected from normal accumulation of the flowstone.

The conclusion of the analysis of the Sulawesi art thus yielded a highly reliable date: the art had to have been affixed to the wall more than 43,900 years ago, with an uncertainty of only a few centuries. This is considerably earlier than any bona fide Western European cave paintings identified to date. The debate over primacy, however, is a rather sterile one. What is clear is that humans, with their uniquely large brains and proclivity for language, engaged in symbolic representations of their world tens of millennia before they settled into fixed camps and launched—10,000 years BP—what we usually regard as "civilization."

STONE TOOLS

Long before they turned to artistic pursuits, our hominid forebears crafted tools, mostly from stone. Debate swirls around (1) what counts as a tool—a few chips off a rock seem to pass muster with most archeologists—and (2) who has found the oldest ones. As this book goes to press, "tools" from a site in Ethiopia hold the record at 2.5 million years.[9] In most of these finds, the dates do not come from the tools themselves but from the geological practice of stratigraphy—noting what layer of rock or volcanic ash the tools are found in and dating that layer via some geological process. One of the helpful techniques for ages in the range of millions of years is the record of magnetic field reversals noted earlier (chapter 8), which leave a faint fingerprint in the rocks themselves.

A more direct and accurate method for determining the age of minerals found associated with stone tools relies on a very rare mode of radioactive decay discussed in chapter 6: spontaneous fission.[10] While several heavy isotopes can in principle undergo spontaneous fission reactions, only ^{238}U does it often enough to leave a signature—and even for this isotope, it occurs less than one in a million times compared to its favored alpha decay. Because the nucleus splits into two unequal pieces, both of which are more tightly bound, a lot of energy is given off, typically around 170 MeV. Most of this appears as the kinetic energy of the two fragments, and they rocket away from each other because of their mutual positive charges repelling each other once they have slipped the surly bonds of the strong force. This displaces atoms and molecules along a straight path until they come to a stop. While these tracks of destruction are only a few nanometers wide, track lengths can extend up to several millimeters.[11]

The tracks are too narrow to see under a light microscope, but by applying chemicals that etch a line along the track, they can be made micrometers wide and thus become easily visible. One simply counts the number of tracks to determine the number of fissions that have occurred. This number is proportional to the number of Uranium atoms in the sample and the time since the crystal was last hot enough to anneal (i.e., heal) the track scars. For zircons, the annealing temperature is about 240°C.

There are two approaches to measuring the amount of ^{238}U present in the sample. The most direct approach is to vaporize part of the sample with a laser and run the material through a mass spectrometer to count the Uranium atoms directly. A second approach is to irradiate the sample with neutrons from a reactor (as in autoradiography) and induce the ^{235}U in the sample to undergo fissions (^{235}U is hundreds of times more likely to undergo fission than ^{238}U). One then counts the fission tracks produced and uses the known ^{235}U/^{238}U ratio (1:138) to infer the number of Uranium atoms present. Dating stone tools from various sites on the island of Flores in Indonesia (recently made famous by the Hobbit-like skeletons found there) M. J. Morwood and colleagues concluded that they are 800,000 to 900,000 years old. This is of interest because it implies that *homo erectus*, the hominid dominant in this period and the likely ancestors of the *homo florensiensis* Hobbits, was capable of crossing miles of open water in order to get to the island.[12]

DATING POOP

The genus *homo*, of which we are members, first diverged from the chimpanzees between 6 and 8 million years ago in Africa. A variety of species branched from this line, some of which left Africa and spread through Europe and Asia, but North and South America remained *homo*-free until quite recently. Our species, *homo sapiens*, arose as much as 300,000 years ago, also in Africa,[13] migrating north and east beginning roughly 70,000 to 150,000 years ago (the precise date is highly controversial). *Homo sapien* teeth from a Bulgarian cave have been ^{14}C dated to 46,000 years ago.[14] Yet even 20,000 years ago, there were no people in the Western Hemisphere.

The exact date for human occupation of the Americas has been a matter of vigorous debate. It is generally agreed that the migration began during the last Ice Age when so much ocean water was piled up in land-based glaciers that the sea level was nearly 400 feet lower than today, creating a grassy land bridge between Siberia and Alaska. Nonetheless, the huge glaciers along the Rockies in Canada and the northwestern United States would have blocked an easy land migration until the glaciers began to retreat at the end of the Ice Age about 13,000 years ago.

For most of the last century, the standard archeological view was that the first humans to occupy North America were members of the Clovis culture, named after the sharpened stone spear points first found in 1929 near Clovis, New Mexico, and subsequently discovered throughout North America and as far south as Venezuela.[15] Traces of animal protein found on some of the points yielded ^{14}C dates of 13,000 years ago, consistent with the dating of the layers of sediment in which they were found.

More recently evidence has been accumulating for an earlier occupation by a so-called pre-Clovis people. In particular, artifacts found in caves in the Pacific Northwest indicated an earlier toolmaking culture that produced duller, chunkier spear points as part of what has been dubbed the Western Stemmed Tradition, which included tools that had similarities to artifacts found in Siberia. Indirect evidence dated these as older than the Clovis dates by 1,000 to 2,000 years. Then, in 2007, Dennis Jenkins and his collaborators found coprolites (preserved feces) in the Paisley caves in Oregon and derived ^{14}C dates greater than 14,000 years.[16] DNA analysis suggested the coprolites were human, but some archeologists questioned whether human handling of the artifacts might have contaminated a sample of animal waste with modern human molecules.

In 2020, Jenkins and his collaborators, led by Lisa-Marie Shillito, analyzed several of those coprolites in which human DNA had been previously detected for lipids, fatty compounds produced by the body that are not soluble in water and thus do not migrate into or out of samples of human remains. They found that several of the coprolite interiors contained unquestionably human lipids, settling conclusively the argument about contamination. One sample had a well-determined carbon date of 14,650 with an uncertainty of less than a century.[17] This is consistent with a scenario gaining credibility in archeological circles in which Siberian migrants began invading the Americas not by land but down the coast of Canada and the United States at least as early as 15,000 years ago. A recent thermoluminescence study of rocks deposited along the coastal route by glaciers suggests they were uncovered by retreating ice as early as 17,000 years ago, leaving the path clear for the seacoast route.[18]

The examples cited in this and the previous two chapters represent but a handful of the thousands of applications that atoms and their isotopes find in reconstructing the creative activities and basic bodily functions engaged in by humans and our ancestors. My personal favorite creative activity centers around food and eating, so before moving beyond anthropocentric subjects, the next chapter will explore the history of human diet and agriculture as revealed by our atomic historians.

CHAPTER 10

You Are What You Eat

tom for atom, it is literally true: you are what you eat. The atoms in your bones and teeth and red blood cells and hair follicles and neurons all once resided in a bagel or a glass of orange juice or a gulp of fresh air, swallowed or breathed in over the course of your lifetime (or, perhaps, over your mother's lifetime, but there are precious few of her atoms remaining in your body today). All those ingested atoms don't remain with you for a lifetime, of course, or you would grow without bound. Some of them are excreted within a day or so after some useful components and/or stored energy has been extracted from them and utilized by your body. Some are incorporated into structures that last more than a lifetime, such as the enamel on your teeth. Others are absorbed into organs or circulating fluids that have turnover times ranging from days to decades.

Whole epithelial cells (from your skin, blood vessels, intestinal linings, etc., each consisting of roughly 300 trillion atoms) are sloughed off in just five days, whereas stromal cells (found, for example, in bone marrow, lymph nodes, and ovaries) are replaced only once every sixteen years on average.[1] At any one moment, your body consists of about 3,000,000,000,000,000,000,000,000,000, or 3,000 trillion trillion, atoms (imagine a global warehouse full of poppy seeds),[2] but they are in a constant state of flux, and what you eat, drink, and breathe determines their relative ratios.

Which atoms are they, exactly? The five most abundant—Oxygen (61.4 percent of your mass), Carbon (22.9 percent), Hydrogen (10 percent), Nitrogen (2.6 percent), and Calcium (1.4 percent)—make up over 98 percent of your total weight. The next ten most common (Phosphorous, Sulfur, Potassium, Sodium, Chlorine, Magnesium, Iron, Fluorine, Zinc, and Silicon) get us to more than

99.99 percent—you have about 1 g of Silicon (0.0014 percent), and everything else is present in even smaller trace amounts. Remember, however, that 0.0014 percent of 3,000 trillion trillion is still a huge number (40 billion trillion to be exact); even the sixtieth most abundant element, Tungsten, includes about 850 million trillion atoms, so there are plenty of them to count. And don't forget, most elements have two or more isotopes that may be present. It turns out your isotope ratios can be used to provide a detailed account of what you eat. Thus, they offer us a window into the evolution of agriculture and human diet.

Forty years ago, I read an article on this subject that I found so elegantly simple and yet so clever and exemplary of the scientific approach to the world that it launched my lifelong interest in using isotopes to reconstruct history. The article, entitled "Carbon Isotopes, Photosynthesis, and Archeology," appeared in *American Scientist*, the bimonthly magazine of the Sigma Xi Society.[3] It was written by Nikolaas J. van der Merwe, who, at the time, was a member of the faculty of the department of archeology at the University of Cape Town in his native South Africa. Educated at Yale and subsequently a professor of anthropology at the State University of New York at Binghamton between 1966 and 1974, he had gone back to South Africa to accept the position at Cape Town. In 1988, he returned to the United States to take up the Landon Clay Professorship in archeology, earth, and planetary sciences at Harvard University, returning to Cape Town in 2000 where he is now professor emeritus of natural history. The first section of this chapter recounts his story.

PLANTS ARE WHAT THEY EAT

It is important to note here that you're not special. Plants are what they eat too. Plants—at the base of your food chain no matter what your diet is—are breathing in air and sucking up both water and minerals through their roots to form fructose, cellulose, and the other molecules that comprise plant material. Thus, the isotope ratios in the air and soil are reflected in the isotope ratios in plant matter and ultimately in you. As always, careful measurements are required. The process of photosynthesis by which plants use sunlight to manufacture their molecules actually discriminates against some isotopes in ways that depend on the plant species. The way your body processes the food you eat again transmutes the isotope ratios as the food becomes bone and blood and neurons. Each step in this process of changing isotopic composition—called fractionation—is easily measurable, so keeping track of it all is a straightforward accounting problem. By coupling this to the imperturbable clocks of the radioactive elements you ingest, the isotopic inventory of fossilized bones can provide us with a direct way to reconstruct the history of human food choices and of the hunting and farming practices on which humans depend.

As a graduate student, van der Merwe had worked in the Yale Radiocarbon Laboratory, where he had the striking insight that one could use radiocarbon techniques to date the origin of Iron tools and other archeological artifacts. The process of smelting Iron from its ore began at least as early 1800 BC in the Hittite Empire and independently in China by 600 BC. The process requires heating the ore to well over 1,200°C which in turn requires the use of charcoal. In this process, Carbon from the charcoal used to heat the ore diffuses into the molten Iron and comprises anywhere from 0.05 percent to 5 percent of the finished product, depending on the temperature achieved. These Carbon atoms (specifically the radioactive ^{14}C) can be used to date the Iron.

It turns out that the best charcoal is made from recently cut green wood—a letter from Hammurabi, king of Babylon in 1750 BC, instructs his servant organizing the charcoal brigade "cut only green wood."[4] Thus, assuming Hammurabi's instructions were followed, the zero-point in time for the tree that made the charcoal is certain—it was cut immediately before diffusing into the Iron.

Van der Merwe demonstrated his ability to date ancient Iron artifacts successfully by determining a date (100 ± 80 AD) for 230 grams of Iron nails from a Roman fort in Scotland known to have been built under the Emperor Agricola in 83 AD and for cast Iron fragments from a tomb in Hohan, China (430 ± 80 BC), known to have been from the Warring States period of Chinese history, which lasted from 480 to 221 BC.

By the 1970s, van der Merwe had decided to include the use of the stable Carbon isotope, ^{13}C, alongside ^{14}C dating in his archeological studies. As we shall see, this heavier, stable cousin of the most common isotope ^{12}C plays a vital role. Van der Merwe summarized the results of his Carbon isotope studies in the paper that stimulated my abiding interest.

Let's begin this story with what plants do. Their "goal" in life is to make plant material, including molecules such as glucose ($C_6H_{12}O_6$), sucrose ($C_{12}H_{22}O_{11}$), and amylose ($C_6H_{10}O_5$), and then link these together in long chains of cellulose, lignin, and other molecules that form the principal structural components of a plant's cell walls. It turns out that plants utilize different photosynthetic pathways to make these molecules, but the fundamental principles are the same. The leaf breathes in CO_2 from the air and, in the presence of H_2O, uses energy from sunlight to break the CO_2 bond, releasing the O_2 (the stuff we breathe) back to the atmosphere and incorporating the C into the organic molecules it is building. To get to even the simplest molecule, glucose ($C_6H_{12}O_6$) takes many steps, but it is the first step in the process that isotopically distinguishes the different mechanisms that plants employ.

C3 VERSUS C4

Greater than 90 percent of all plant species follow the so-called C3 pathway that links together three Carbon atoms from three different CO_2 molecules to start the chain. This pathway evolved several billion years ago when the CO_2 concentration in the atmosphere was much higher than it is today and the O_2 concentration was lower. About 30 million years ago, when the CO_2 concentration had fallen significantly (and billions of years of photosynthesis had raised the O_2 levels to modern values), a more efficient pathway, C4, evolved, and its first step linked not three but four Carbon atoms. This turned out to be particularly advantageous in hotter and drier climates, so while such species make up fewer than 3 percent of all plants alive on Earth today, C4 plants are responsible for the production of roughly 25 percent of global plant material, and a full 46 percent of agricultural grain production because corn, sorghum, and millet all use the C4 pathway.[5] A third mechanism, also a late arrival, evolutionarily speaking, is named CAM (for the tongue twister of the day: crassulacean acid mechanism); it provides the first step of photosynthesis in the other 6 percent of plant species (primarily cacti and succulents, including important contributors to tropical umbrella drinks such as *agave tequilana* and pineapples).

Now, the CO_2 molecules in the air reflect the ratios of Carbon isotopes that we laid out in chapter 4: 98.9 percent contain ^{12}C, 1.1 percent are ^{13}C, and one in a trillion are ^{14}C. Recall that all isotopes of an element are chemically identical; that is, a $^{12}CO_2$ molecule is chemically indistinguishable from a $^{13}CO_2$ molecule, and its Carbon atom will behave the same way when this molecule is broken apart by a packet of solar energy and subsequently used in building sucrose or whatever. The one distinguishing feature of the $^{13}CO_2$ molecule, however, is that it is heavier than its $^{12}CO_2$ cousin and thus, as this former marathoner who has gained a lot of kilos later in life can tell you, heavy things move more slowly. The chubby $^{13}CO_2$ molecules just aren't as nimble as their lighter brethren.

As a consequence, when a C3 plant starts scavenging CO_2 to begin the photosynthetic process, it more readily finds the faster $^{12}CO_2$ molecules—they get around, while the slovenly, heavier isotopes show up less frequently. The result is that C3 plant material has, on average, 1.95 percent less ^{13}C than was present in the air from which it came.[6] C4 plants, on the other hand, have to be more patient because they need to collect four CO_2 molecules and split them before the first step in the process is complete. While they too tend to abjure the slower $^{13}CO_2$ molecules, they can't afford to be quite so picky and, as a result, their cellulose is deficient in ^{13}C by only about 0.55 percent[7] (see figure 10.1).

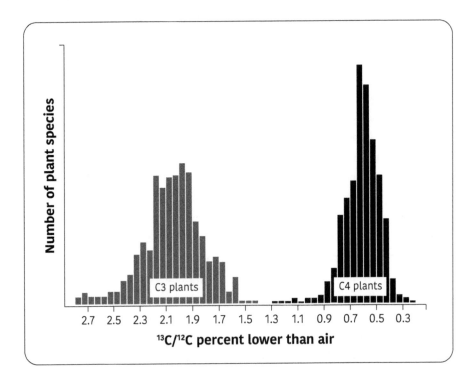

10.1 The distribution of the $^{13}C/^{12}C$ ratios for C3 (light gray) and C4 (dark gray) plant species. The distributions do not overlap and peak at roughly −2.0 percent and −0.5 percent, respectively, compared to the ^{13}C content of the air. The negative values indicate that, as described in the text, all plants discriminate against the heavier, slower-moving ^{13}C isotope, but the shorter C3 photosynthetic pathway discriminates to a greater degree.

Returning to this chapter's theme—you are what you eat—we should expect to find different $^{13}C/^{12}C$ isotope ratios in your body depending on whether you eat C3 or C4 plants. However, the process of digesting food and turning it into body tissue is also a chemical process, each step of which suffers from similar differential discrimination both for and against heavier atoms. In fact, turning plant material into bone collagen ups the original ratio by +0.51 percent, making muscle boosts it by +0.36 percent, and making fat decreases it by −0.30 percent. As noted above, these chemical discrimination values are called fractionation factors. Thus, the bone of a committed C3 plant eater will have a $^{13}C/^{12}C$ ratio smaller than the air by −1.95 percent + 0.51 percent = −1.44 percent, while a pure C4 consumer will have a ratio of −0.55 percent + 0.51 percent = −0.04 percent, which is essentially zero, the same ratio as the air from which the Carbon ultimately came.

If you aren't a vegetarian, things get a little more complicated, but the principles remain the same. Meat is muscle, and if you eat a lamb chop from a sheep grazing on C3 plants, your ingested $^{13}C/^{12}C$ ratio would just be −1.95 percent (from the plant) + 0.36 percent (from conversion to muscle in the sheep) + 0.51 percent (from conversion to bone in you) = −1.08 percent compared to the air.

These small percentages may seem like such trivial differences that they couldn't possibly be measured but, remember, atoms are *tiny*. With even 1 gram (0.04 ounces) of bone, there would be a million trillion fewer ^{13}C atoms[8] than expected—not a small discrepancy at all. With our ability to deconstruct sample materials and separate them into their atomic—and isotopic—constituents, these seemingly small discrepancies become powerful tools to unlocking the secrets of the past.

THE STORY OF MAIZE

Maize—what we more generally refer to as corn today in the United States—originated in the highlands of southern Mexico. A consensus view is emerging that it was derived from teosinte, a native flowering grass endemic to this region. Teosinte looks quite different from a modern corn plant, with its handful of seeds (kernels) clustered on a one-inch stalk, but it has striking genetic similarities with corn. Many authors suggest that selective breeding of teosinte, begun as early as 9,000 years ago by local hunter-gatherers, led to maize cultivation before

4000 BC; indeed, the oldest known corn cobs, from the Guila Naquitz cave in Oaxaca, Mexico, have been Carbon-14 dated to 4235 BC, 6,250 years ago.[9]

By the time Europeans arrived in the New World, corn was a principal staple of the Indigenous peoples of North and South America and the Caribbean. Today, it is by far the leading grain crop in the world, accounting for 1,100 million metric tons in the 2018–2019 harvest, or roughly 43 percent of all global grain production.[10] How did cultivation of this food crop spread from an obscure mountain valley in southern Mexico to become the world's most popular grain? Carbon isotopes hold the key.

Nearly all the natural vegetation covering North and South America, both trees and grasses, as well as blueberry bushes, potatoes, tomatoes, cassava, manioc (more on manioc later), and other food crops native to the Americas are C3 plants. Corn is a C4 plant, the only edible C4 plant native to the New World. Thus, in principle, we should be able to tell from the bones of the Indigenous people when this new crop was introduced to their diet.

And this is precisely what van der Merwe and his collaborator J. C. Vogel managed to do. Analyzing over fifty skeletons discovered in Illinois, Ohio, and West Virginia, they used Carbon-14 dating to establish ages ranging from 3000 BC to 1300 AD. For all those remains older than 500 AD, the $^{13}C/^{12}C$ isotope ratios clustered around −1.45 percent—the value expected from a diet of exclusively C3 plants, with an average ratio of −1.95 percent added to the +0.5 percent fractionation factor of bone collagen.

For the skeletons dated to between 600 and 1200 AD, however, the $^{13}C/^{12}C$ ratios show a steady rise, reaching an average value of −0.45 percent, a full 1 percent increase, by the end of this period (see figure 10.2).[11] It does not reach the value of zero expected from a fully C4 plant diet because, even if you now have this delicious new source of protein, starch, and sugar, it doesn't mean you must give up eating blueberries and tomatoes and the occasional venison chop. The measured ratio allows us to calculate the fraction of C4 plants in the diet of Indigenous Ohioans from 5,000 to 800 years ago. Remarkably, over the space of just a few centuries, it goes from 0 percent (the era before corn) to over 70 percent.

While van der Merwe was teaching in Binghamton, New York, he had the opportunity to repeat this experiment on a series of human ribs excavated from sites in upstate New York and dated from 2500 to 100 BC, plus a second set from 1000 to 1500 AD. Again, the earlier bones had the expected ratio of −1.4 percent

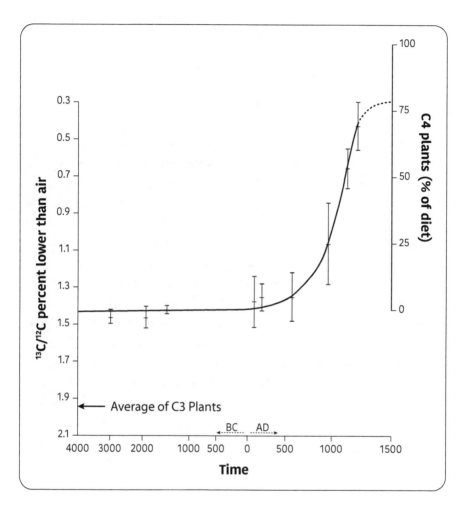

10.2 The ratio of $^{13}C/^{12}C$ in the bones of Indigenous Americans from the midwestern United States as a function of time over the period 4000 BC to 1500 AD. Note the rapid rise in the skeletons dating from 1000 AD. The scale on the right shows the fraction of diet that comes from C4 plants. This fraction is essentially zero for the first 5,000 years and then, within a few centuries, rises to over 70 percent, indicating the cultivation of maize.

for a largely C3 diet with a little animal meat thrown in, while the latter group showed an increase over time from −0.95 to −0.65, showing a growing reliance on C4 corn, albeit slightly less than in Ohio, possibly because of the somewhat shorter growing season in upstate New York. This work has now been repeated for sites throughout North and South America, allowing us to map the spread of corn cultivation from a mountain valley in Oaxaca to becoming the dominant farmed crop throughout the New World.

MORE YOU-ARE-WHAT-YOU-EAT STORIES

Anna Roosevelt obtained her PhD from the department of anthropology at Columbia University in 1977, the year I arrived at Columbia. Within a few years, she had sparked a controversy in Amazonian archeology based on her excavations in the Orinoco valley of Venezuela. The standard view at the time was that maize had played no role in the civilizations of the rain forests of South America in prehistoric times (1,500 to 3,000 years ago). Rather, a system known as the "tropical rain forest system," based on the cultivation of manioc and cassava for starches and supplemented by hunting and fishing for protein, was said to have supported the populations there.[12]

Roosevelt's excavations suggested, however, that the large chiefdoms of this period, with their high population densities, could not be supported by the tropical rain forest system alone because (1) animals were too scarce, (2) fishing was impossible in the rainy season when the rivers were too turbulent, and (3) root crops would rot in the wet part of the year. Instead, she proposed that maize was introduced as much as 2,500 years ago and supplied the protein needed when hunting and fishing were insufficient. Her views were stoutly resisted by the scions of the field.

This might be seen as the kind of academic argument that only evolves as the protagonists die off, but Carbon isotopes came to the rescue. Most of the tropical vegetation was C3 plants, and analysis of animal bones from the period showed that the few C4 grasses that they could, in principle, have eaten to lower their $^{13}C/^{12}C$ ratios were not relevant—the turtles, fish, and the large rodents the Indigenous people ate all fed principally on the C3 plant material.

There remained a critical question, however. Was the cultivated manioc a C3 or a C4 plant? This was settled by a seven-hour round-trip sojourn from Binghamton to a Puerto Rican market in New York City to buy two manioc cakes. The $^{13}C/^{12}C$ ratio was −1.9 percent—as expected for a standard C3 plant. However, two skeletons from the banks of the Orinoco River at Parmana, several hundred miles upriver from the Caribbean, dated to 2,800 years ago, yielded a surprising result: the bones had a $^{13}C/^{12}C$ ratio of −1.9 percent—the same as pure

C3 plants without any of the +0.5 percent addition expected from the bone formation process.

Another isotope measurement by E. Medina and P. E. H. Minchin solved this riddle.[13] They showed that, near the ground in a tropical forest, the air had a $^{13}C/^{12}C$ ratio that was 0.5 percent *lower* than at the tops of the trees because the air trapped near the ground by the dense canopy is enriched by rotting plant material that itself was deficient in ^{13}C.[14] This extra half percent just canceled the +0.5 percent of the bone fractionation factor, yielding the expected result for a heavily C3 diet. This is a prime example of the care that must be taken when querying the isotope record. As noted in the introduction, atoms may escape the cultural biases of human historians, but they are subject to a variety of physical and chemical effects, all of which must be carefully assessed before drawing definitive conclusions.

And what about bones from a later period, roughly 400 AD, when the major chiefdoms were thriving? They gave an average value of −0.33 percent, even lower than that of their corn-loving North American cousins, indicating a diet made of a roughly 80 percent maize. *Quod erat demonstrandum*[15]—Roosevelt's hypothesis was confirmed.

OTHER ISOTOPES AND HUMAN DIET

It is the nature of biological processes to build complex molecules from simple building blocks. As we have seen with Carbon, this construction process leads to biased isotope ratios that provide us with clues to the origin of these building blocks. But Carbon isn't the only element for which this is true. For example, Nitrogen—the next element in the Periodic Table and the fourth most abundant one in your body—also has two stable isotopes, ^{14}N and ^{15}N. As with Carbon, the heavier isotope is much rarer, making up just 0.64 percent of the Nitrogen found in our atmosphere, where N_2 is the dominant component (comprising nearly 78 percent of the air). While many biological molecules require Nitrogen atoms, neither plants nor animals can use the abundant atmospheric N_2 because the bond between the two atoms is too hard to break. To become biologically useful, the Nitrogen must be "fixed" (i.e., broken apart and incorporated into other molecules that biological systems can deconstruct and use). This task falls to several species of bacteria collectively labeled diazotrophs.[16] Some of these bacteria live symbiotically on the roots of certain plants called legumes (peas, beans, clover, alfalfa, peanuts, etc.), where they actively convert the useless atmospheric N_2 into usable forms of Nitrogen to help those plants outcompete their neighbors. When legumes die (or are plowed under), the fixed Nitrogen is released into the soil, where it can fertilize the growth of plants lacking these friendly bacteria.

Cyanobacteria, the original blue-green algae that transformed the Earth by injecting vast quantities of Oxygen into our atmosphere, are also great Nitrogen fixers, especially in the ocean. While the legume bacteria can fix up to 90 kg of Nitrogen per acre per year in a field of clover, cyanobacteria on coral reefs can fix up to 250 kg per acre per year (a difference that we shall discover later in this chapter to be important).

Unlike the case of Carbon, most plants don't discriminate much against the heavier isotope of Nitrogen. Most plant material lies well within ±0.5 percent of the value found in the air if grown using humanmade fertilizers (synthetic fertilizers use N_2 from the air in their manufacture). However, natural fertilizers (e.g., manure) do have higher ^{15}N values that can bias plant ratios to the positive side of zero. Indeed, this distinction can be used to identify truly organically produced crops, as discussed below.

Herbivores, including all the animals we grow for meat, incorporate the Nitrogen atoms from their food into their protein molecules. However, animals also continuously excrete compounds that include Nitrogen; for example, the urea molecule, with the chemical formula $CO(NH_2)_2$, contains two Nitrogen atoms each. This excretion process favors the lighter ^{14}N isotope. Thus, animal tissues and fluids concentrate the heavier ^{15}N isotope by about 0.3 to 0.4 percent over that found in the food they eat.

Thus, herbivores have average values of $^{15}N/^{14}N$ of +0.5 percent. Carnivores (which typically eat herbivores) represent a second step of concentration of the ^{15}N isotope because they also preferentially excrete the lighter form of the atom, yielding an average ratio of +0.8 percent. An exception to this rule is found in the ocean, where the food chain has several more steps than on land, from the plankton at the base all the way up to the carnivorous fish and marine mammals we eat. Beginning with phytoplankton, with $^{15}N/^{14}N$ values of +0.5 to +0.7 percent, we climb the chain to the zooplankton that feed on them (+0.8 to +1.0 percent) to the small fish that eat the zooplankton (+1.1 to +1.3 percent), to the larger carnivorous fish we eat such as salmon and tuna (+1.4 to +1.7 percent), to the seals and orcas that eat the big fish (+1.8 to +2.0 percent).[17]

This discrepancy between seafood versus land-based plants and animals allows us to reconstruct the diets of prehistoric peoples. M. J. Shoeninger, M. J.

DeNiro, and H. Tauber found coastal dwellers ranging from California to Denmark in Neolithic times had ^{15}N/^{14}N ratios of +1.5 to +2.0 percent in their bones; accounting for the extra step of concentration in these humans (who also excrete urea) suggests that a large fraction of their diet came from fish, with a modest admixture of plants (and possibly birds or mammals).[18] In contrast, both Mexican and European agriculturalists had Nitrogen isotope ratios of +0.8 to +1.1 percent, with ^{13}C/^{12}C ratios for the former lying near −0.7 percent (from C4 corn) and the latter at −1.8 to −2.1 percent (from their C3 plant diet).

These authors did find one puzzling result for natives of the Bahamas: their ^{15}N/^{14}N ranged from +1.0 to +1.3 instead of nearly +2.0 percent, despite the obvious abundance of local fish and the paucity of land on which to grow crops. As with the Carbon from the rain forest floor, however, we must always be alert for confounding factors. As noted above, coral reefs are dense with cyanobacteria busily fixing Nitrogen, which depletes the ^{15}N more readily in the surrounding water, leading to fish (and ultimately bones) that are closer to the Nitrogen isotope ratio of herbivores on land.

Sulfur is yet another element essential for life (the seventh most common in the body at 0.2 percent) and has four stable isotopes. The dominant isotope is ^{32}S, but 4.25 percent of Sulfur is in the heavier isotope ^{34}S (the other two, ^{33}S and ^{36}S, are both well under 1 percent). The two dominant forms of Sulfur on the Earth's surface are Sulfur from igneous (volcanic) rocks and the sulfate molecule SO$_4$ found in the ocean. The ^{34}S/^{32}S ratio of the former, analogous to the Nitrogen ratio in the air, defines the zero point, while sulfate in seawater has a ratio higher by +2.1 percent. Unlike Nitrogen, Sulfur is not excreted by animals, so the Sulfur isotope ratios throughout the food chain are similar to those of the ultimate source (either rocks/soil or seawater). Thus, the Sulfur isotope ratio can be used as a secondary indicator of the amount of seafood in a diet, independent of the Nitrogen anomaly found for reef fish.

MODERN USES OF ISOTOPE RATIOS

At the risk of sounding like a broken record, we *still* are what we eat. In the last fifteen years, stable isotope ratios of Carbon, Nitrogen, and Sulfur have begun to be employed in the studies of human diet and health. An early study by B. Buchardt, V. Bunch, and P. Helin in 2007 compared the isotope ratios in the fingernails collected from an Inuit community in northern Greenland (eighty-two samples) and a group living in Denmark, including a few Greenlanders who had relocated there.[19] The isotope ratios of all three elements were highly correlated with each other and markedly different between the two populations (see figure 10.3).

As expected, the Inuit, with a diet rich in fish and marine mammals, had high values of all three ratios, whereas those living in Denmark (whether Danes or

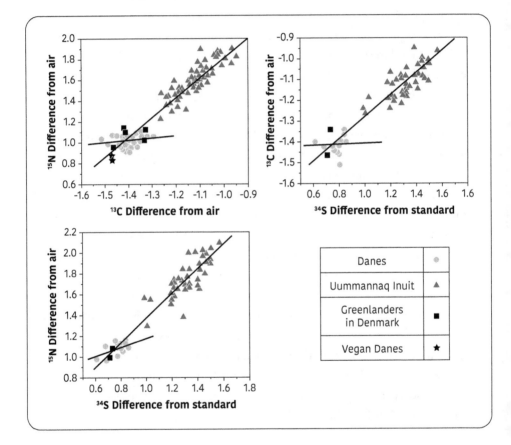

10.3 The correlations of ^{13}C versus ^{15}N, ^{34}S versus ^{13}C, and ^{34}S versus ^{15}N for populations of Inuit from northern Greenland, Greenlanders who had moved to Denmark, native Danes, and two vegan Danes derived by Buchardt et al. (2007) from fingernail samples. Note how the Greenlanders who had moved to Denmark overlap completely with the native Danes, showing the dominant influence of diet. For example, the Greenlanders who eat mostly fish and carnivorous marine mammals have high values of ^{15}N and ^{13}C, while the vegan Danes have among the lowest ^{13}C and ^{15}N, as expected in a plant-based diet. Recall that plants discriminate against the heavier isotopes and animals concentrate the heavier ^{15}N in their bodies.

Greenlanders) had much lower values based on a diet richer in vegetables and herbivores. The two Danish vegans had the lowest Nitrogen ratios of all, reflecting a purely plant diet with no animal concentration of the heavier isotope.

Researchers have even demonstrated how dietary interventions work in real time to set the body's isotope ratios. By switching four volunteers, two men and two women, from their normal diets dominated by C3 plants and animal protein to a diet rich in C4 plants and marine-based protein for twenty-eight days, they found distinct changes along strands of the subjects' hair as the new diet reset the isotope ratios: the $^{13}C/^{12}C$ ratio changed by +0.85 to +0.90 percent, and the $^{15}N/^{14}N$ ratio changed by +0.15 to +0.22 percent along a single strand of hair as it grew over the course of the month.

An interesting example of the sensitivity of isotope ratios to diet and the condition of the organism is provided by the change in the Nitrogen isotope ratio that occurs in cases where food (particularly protein) intake is insufficient. This effect was first noted in animal experiments—newly hatched quail fed a restricted diet had $^{15}N/^{14}N$ ratios +0.19 percent higher than their siblings who were allowed to eat as much as they desired,[20] a consequence of the food-restricted body breaking down its own proteins and excreting wastes that are enriched in the light isotope. This effect has subsequently been observed in anorexics who enter recovery programs: hair samples show a systematic gradient from the tips (from the pre-entry, "sickest" period) to the roots ("healthiest" time prior to release from the recovery program).

A striking example comes from the monthly sampling of a woman who became pregnant and suffered severe morning sickness in the first trimester, leading to a weight loss of 12 pounds, followed by her recovery and a weight gain of 18 pounds above her original weight at the time of the birth. During the period of morning sickness, her Nitrogen isotope ratio spiked up by a full +0.10 percent. This was followed by a decline of −0.16 percent as her weight recovered when she reached full term (see figure 10.4).[21]

As more data have accumulated, some scientists have started to look at isotope ratios as tools for studying both nutrition and disease. For example, using diets with different levels of C4 sugars (both high fructose corn syrup and cane sugar are C4), C. M. Cook and colleagues found corresponding changes in the $^{13}C/^{12}C$ ratio in blood glucose after a meal.[22] P. S. Patel and colleagues found an inverse correlation between fish intake, indicated by low Carbon isotope ratios, and diabetes risk.[23] The liver is the site of urea production and protein synthesis, both of

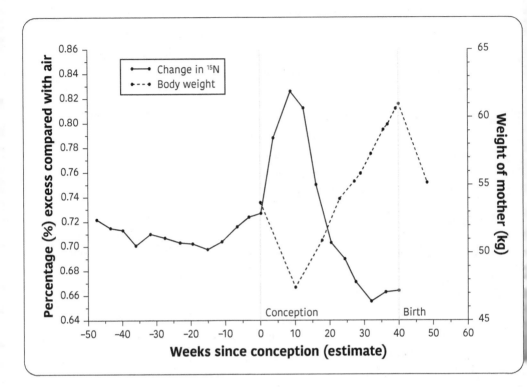

10.4 Measurements of the ratio of $^{15}N/^{14}N$ compared to the ratio in air as a function of time for a woman who became pregnant (conception equals week 0). Also plotted is the woman's weight. Morning sickness in the first trimester caused significant weight loss and a contemporaneous spike in the Nitrogen ratio, which reversed as soon as weight gain commenced (from Reitsema 2013).

which fractionate Nitrogen. K. J. Petzke and colleagues found that patients with cirrhosis—a degenerative liver disease—had Nitrogen isotope ratios in their hair that was −0.32 percent lower than healthy subjects.[24]

The heavy isotopes of Hydrogen (^2H currently makes up about 0.02 percent of all Hydrogen) and Oxygen (^{18}O with an abundance of about 0.25 percent that of ^{16}O), when present in water (H_2O), make the water "heavy." In addition to providing a sensitive proxy for both temperature and an indicator of rainfall location (see below and chapter 11), the fractionation involved in water produced from metabolic reactions leads to systematically higher heavy water fractions than those found in water ingested in food and drink. Because diabetes causes dehydration and thus the ingestion of more drinking water, mice with induced diabetes have been shown to have $^2H/^1H$ and $^{18}O/^{16}O$ ratios that are −0.13 percent and −0.27 percent lower, respectively, than their healthy cousins. While this effect has not yet been established in human diabetics, T.-C. Kuo and colleagues have shown lighter isotope ratios in patients with kidney disease.[25]

Calcium has five stable isotopes, of which ^{40}Ca is dominant (96.9 percent), and ^{44}Ca makes up most of the rest (2.1 percent). Fractionation processes lead to human bone having an excess of the lighter isotope compared to the ratio in food, whereas both soft tissues and urine contain more of the heavy isotope. In osteoporosis, bone minerals are resorbed and excreted in the urine which thus has more of the light isotope than normal. Urinalysis thus provides a simple, noninvasive way to monitor bone loss.

Finally, isotope ratios can also be employed in forensic investigation of food labeling. As noted above, organic crops are supposed to be grown using only natural fertilizer (i.e., animal waste), which, because such waste is higher in heavy ^{15}N, should have higher $^{15}N/^{14}N$ values than crops grown using humanmade fertilizers. Pure clover honey should have a C3 plant $^{13}C/^{12}C$ ratio; any departure from this suggests adulteration with (much cheaper) high fructose corn syrup or cane sugar. Likewise, brandy, distilled from C3 grapes, should have no signature of fortification with distilled grain alcohol.

ÖTZI'S JOURNEY: THE POWER OF ISOTOPES

While hiking in the Alps near the Italian-Austrian border in September 1991, two German tourists came across a human torso visible under the edge of a retreating glacier. Assuming it was a deceased hiker, possibly missing for years, they notified authorities. In a sense, they were correct, but when the body was recovered and dated using C-14, it was discovered that Ötzi, as he came to be called, met his demise while hiking 5,200 ± 70 years ago.[26] Detailed analysis of this partially mummified body has led to fascinating insights regarding the clothing, tools, and diets of Copper Age Europeans. Of relevance to our current topic of isotopes, however, is the remarkable precision with which we can identify his birthplace and subsequent travels.

As noted in the introduction to this chapter, the longest lasting atoms in the human body are found in tooth enamel. The layers of this hard substance called apatite are made of crystalline calcium phosphate (primarily $Ca_5(PO_4)_3(OH)$). They are deposited in layers as the permanent tooth grows in, starting from birth until the age of approximately five years.

The element Strontium lies directly under Calcium in the Periodic Table of the Elements and thus, as discussed in chapter 4, has very similar chemical properties to Calcium and is easily substituted for it in the apatite crystals. In contrast to tooth enamel, human bones have a turnover time for their atoms of ten or twenty years. Thus, enamel provides evidence of where a subject was born, while the bones contain a signature of where they spent their last decade or two of adult life. Ötzi, for instance, is estimated to have been about forty-five years old at the time of his death.

Wolfgang Muller and his colleagues examined the four distinct geological provinces of northeast Italy and characterized them by their isotope ratios of Strontium and Lead: $^{87}Sr/^{86}Sr$ and $^{206}Pb/^{204}Pb$, respectively. These soil minerals are taken up by local plants that, when eaten, are incorporated into teeth, thus reflecting Ötzi's birthplace. The result of an analysis of his canine tooth enamel (with a $^{87}Sr/^{86}Sr$ ratio of 0.7215) rules out two of the geological provinces.

The researchers then examined the $^{18}O/^{16}O$ ratio in the enamel. In clouds, the heavier ^{18}O-containing water rains out first, leading to a decreasing Oxygen isotope ratio as one moves farther from the source of evaporation (in this case the Mediterranean Sea). The tooth enamel indicated that Ötzi was drinking water with a ratio of −1.1 percent compared to the standard value when he was three to five years old. This finding rules out the southern geological provinces (nearest the ocean), which have values substantially greater, pinpointing Ötzi's birth to the vicinity of the Eisack Valley about 15 miles north of modern Bolzano.

Analysis of Ötzi's femur bones shows a considerably different Strontium $^{87}Sr/^{86}Sr$ signature from his teeth: $^{87}Sr/^{86}Sr$ = 0.7177. In addition, an analysis of small chips of mica found in his stomach, inferred to be there from accompanying the milled grain he had recently eaten, have an Argon ratio $^{40}Ar/^{39}Ar$ that also suggests he had moved roughly 15 miles west of his birthplace to near modern-day Merano. His final resting place was another 15 miles northwest on the Austrian border. From tiny atoms, the story of Ötzi's lifetime can be told 5,000 years after his death.[27]

Isotopic ratios apply equally well to creatures other than humans, allowing us to recount their life histories as well. Recently, M. J. Wooller and colleagues reconstructed the life of a wooly mammoth that lived in Alaska 17,100 years ago before the first humans reached North America.[28] By analyzing ratios of $^{87}Sr/^{86}Sr$, $^{18}O/^{16}O$, $^{15}N/^{14}N$, and $^{13}C/^{12}C$ recorded along the 8-foot length of the mammoth's tusk, they could reconstruct the animal's location at *weekly* intervals over its entire twenty-eight-year lifetime. The Strontium and Oxygen ratios in the tusk mirrored those in the plants the mammoth ate, which in turn reflected the underlying geology of the location.

DNA revealed a single X-chromosome, indicating a male specimen. In its first eighteen months, it lived near the Yukon River, ranging from the Seward Peninsula to the Yukon Delta. It spent the next fourteen years traversing over 1,000 km mostly south of the Brooks Range. Fifteen years is about the age when young males are expelled from the matriarchal herd (in modern-day elephants, at least) and take off on their own. Our mammoth expanded its range. When it was nineteen, twenty, twenty-two, and twenty-three (but not twenty-one, twenty-four, and twenty-five) it showed a pattern of plunging Sr ratios once a year, indicating seasonal migration. It spent its final two years north of the Brooks Range; a spike in its $^{15}N/^{14}N$ ratio at the end of its life indicated that it probably died of starvation (recall the pregnant women with morning sickness).

A similar study of a 13,200-year-old mastodon from Indiana also shows a marked change in range at age twelve to fourteen, when it would have left its natal herd. Using the temperature-dependent $^{18}O/^{16}O$ ratio, J. H. Miller and colleagues infer that the animal returned to northeast Indiana each year in the spring or early summer, and for at least its last eight years, it had battle scars on its tusk, presumably from clashes over mates; it died from a puncture wound to the head from its final clash before its thirty-fifth birthday.[29] Studies such as these open the way for a detailed reconstruction of the lives of North American large fauna as they plunged toward extinction at the end of the last Ice Age when humans invaded the continent.

The changes in isotope ratios we have been discussing are small, typically measured in tenths of a percent or in parts per thousand. A value of 0.1 percent is, in most circumstances, small enough to be ignored—if your $850 credit card bill were off by 85 cents, you likely wouldn't notice. When assessing isotopic abundances, however, such a value is both easily measurable and profoundly revelatory, providing us with the otherwise unknowable history of diet and agriculture, as well as offering new insights into human nutrition and disease and the life histories of our forebears and their environments.

CHAPTER 11

Paleoclimate

Taking the Earth's Temperature Long Ago

I n 1975, my late Columbia colleague Wally Broecker[1] published an article in *Science* magazine entitled "Climatic Change: Are We on the Brink of Pronounced Global Warming?"[2] In that year, the integrated global temperature was 0.03°C below the average for the twentieth century; over the previous decade, the average was exactly equal to that long-term mean value. Broecker predicted that those recent trends would "within a decade or so, give way to a pronounced warming induced by carbon dioxide." He went on to predict that "the exponential rise in the atmospheric carbon dioxide content will tend to become a significant factor and by early in the next century will have driven the mean planetary temperature beyond the limits experienced during the last 1000 years."[3] In fact, from 1977, eighteen months after the article appeared, through today, every single year has seen a global temperature higher than the long-term average, with a steady increase of 0.14°C per decade until 2015, after which the warming has accelerated. In 2020, the temperature was a full degree Centigrade (1.8°F) above the twentieth-century average, warmer than the Earth has been in several thousand years.

Given the average conditions that existed in the decade before 1975, it is clear that Broecker's prognostication was not a guess based on an extrapolation of an existing trend but a prediction based on physics. Our focus in this chapter is on determining temperatures in the distant past to see what they portend for our future. Thus, it seems worthwhile to spend a little time exploring what determines the Earth's (or any planet's) temperature.

SETTING THE THERMOSTAT

There is a very simple equation that allows one to calculate the temperature of any object, be it a planet or your sister-in-law:

$$\text{Energy in} = \text{energy out}$$

More than 99.97 percent of the energy arriving at the surface of the Earth comes to us from the Sun.[4] As we discussed in chapters 3 and 4, it is the temperature on the surface of the Sun that provides a measure of how fast the atoms there (ions mostly) are moving, and the speed of those particles governs the wavelengths of light they emit. The Sun's surface temperature is 5780°K, leading to the emission of light at wavelengths of around 500 nanometers, yellow-green light that is right in the middle of the single octave of wavelengths our eyes can see.[5]

Earth's atmosphere is almost completely transparent to these wavelengths, so almost all of this energy gets through except for the fraction reflected back into space by clouds. An additional fraction that reaches the ground also reflects directly into space by bouncing off the ice sheets, deserts, and other shiny surfaces. None of this reflected light contributes to the Earth's energy balance. In total, about 69 percent of the energy from the Sun incident on the upper atmosphere is absorbed by the Earth, with the remaining 31 percent reflected back into space.

If that were all that happened, the Earth would just get hotter and hotter like an oven without a thermostat. To maintain a reasonably steady temperature, the Earth must *radiate* back into space exactly the same amount of energy it absorbs—and it does. However, there is one critical difference in the energy radiated compared to that absorbed. Because the Earth is at a temperature of (on average) 16°C = 289K, it radiates this energy at wavelengths considerably longer that those received from the 5780K Sun; because it is about twenty times cooler, the peak outgoing light is in the infrared part of the spectrum, with an average wavelength twenty times longer, or about 10,000 nanometers. And the atmosphere is *not* transparent to all those wavelengths. In particular, molecules of water vapor (H_2O), carbon dioxide (CO_2), ozone (O_3), methane (CH_4), nitrous oxide (N_2O), and others in the air absorb some of the outgoing wavelengths and radiate them back toward the Earth's surface.

Again, if that were all that happened, we'd be in an oven. But the presence of this layer of absorbing gases throwing back extra energy causes the surface to warm (recall that energy must be conserved, it can't just disappear). This in turn causes the molecules of the ocean and the land to vibrate a little more vigorously, raising the temperature and shortening the wavelengths emitted until they are short enough to sneak through and escape into space. This phenomenon is called the greenhouse effect because it is analogous to how the glass of a greenhouse lets the sunlight in but is partially opaque to the outgoing infrared rays and thus heats up the air inside the greenhouse. If it weren't for our blanketing atmosphere, the average temperature of Earth would be about −5°C (23°F), and the planet would likely be a frozen world on which life may well have never arisen. The greenhouse effect, like a blanket on a chilly evening, is a good thing.

It is possible, however, to have too much of a good thing. Five blankets on a chilly evening are likely to have you waking up after an hour or so bathed in sweat—you've now blocked *all* the heat in the form of infrared radiation trying to leave your body and, because your digestive system is blithely working away, continuing to release the energy in that ice cream you had for dessert, your temperature starts to rise.

The global warming Broecker predicted is exactly this type of phenomenon. Carbon dioxide is released into the atmosphere when the Carbon atoms in fossil fuels (long-dead plants that sucked CO_2 out of the air 100 million years ago or more in the age of the dinosaurs) are burned (i.e., combined with O_2) to release the CO_2 again. And the increasing number of CO_2 molecules in the air are acting like the extra blankets. Since the beginning of the Industrial Revolution in the late eighteenth century, we have dumped just about 1,000 billion tons of CO_2 into the atmosphere. Thus, just as Broecker predicted, the Earth's surface temperature is rising to levels not seen for thousands of years.

While a temperature change of 1°C may not seem terribly significant, the effects of this change are already becoming apparent in more intense and protracted heat waves, forest fires from the Amazon to Siberia, more violent storms, melting glaciers, rising sea levels, and shifting patterns of drought that could well disrupt the agricultural production needed to feed a world of 8 billion people. As noted above, the rise in temperature appears to be accelerating, and our best current models predict that if we continue along our current path, the mean global temperature rise will be at least 4°C (more than 7°F) by the end of this century. This predicted temperature rise would cause colossal global disruption, including the forced migration of hundreds of millions of people; massive food shortages; sea-level rise of a meter or more, which would flood many of the world's coastal cities; and other deleterious impacts. How confident are we of these predictions?

A detailed assessment of Earth's climate and the models used to predict its future evolution lie outside the scope of this book. The climate system is complex, and many interacting parameters need to be carefully considered. A major limitation in building climate models is the relative paucity of good data we have with which to test them. The directly measured temperature record extends back at most 140 years, and records of precipitation, drought occurrence, atmospheric composition, and other important factors are even less complete. It would clearly be helpful to have longer and more comprehensive records of past climate against which to compare our models for the future. Our atomic historians fortunately provide a wealth of such data extending back millions of years. Their story will be the subject of this chapter. But we begin by addressing the question, Where is all the extra CO_2 showing up in the atmosphere coming from?

WHENCE ALL THE NEW CO_2?

Direct measurements of the CO_2 content of the atmosphere began in 1958.[6] Over the past sixty-five years, the number of CO_2 molecules has gone from 315 for every million particles of air to 420 parts per million (ppm), a 33 percent increase.[7] We know from some of the records, discussed later in this chapter, that such variations have occurred in the distant past, long before humans existed— the CO_2 content has been both much higher and much lower than it is today. So how can we be sure that it is human activity, specifically the burning of fossil fuels (and of tropical rain forests), that has caused the recent rise? Might this not be a fluctuation caused by the same natural mechanisms that have led to large changes in the past? As is becoming routine in our historical excursions, Carbon isotopes provide the definitive evidence.

We saw in chapter 10 that plants are what they eat, and most of what they eat is CO_2. But in examining the details of photosynthesis, we found that they are not unbiased eaters—they discriminate against the slower-moving heavier isotopes ^{13}C and ^{14}C when they choose CO_2 molecules from the air to incorporate into their plant molecules. We also saw in chapter 8 how trees (mostly C3 plants) provide a perfect calendar for studying the past, with each growth ring corresponding to a specific year and our total record extending back nearly 14,000 years. We used the ratio of $^{14}C/^{12}C$ in individual tree rings to calibrate the production rate of Carbon's radioactive isotope to improve the accuracy of C-14 dating. Here, we use the stable isotope ratio $^{13}C/^{12}C$ along with the $^{14}C/^{12}C$ ratio to identify unambiguously the source of the atmosphere's rising concentration of CO_2.

Because ^{13}C and ^{12}C are both stable isotopes, and the total amount of each in the environment is constant, we should expect the $^{13}C/^{12}C$ ratio in an old piece of wood to be the same as it is in a living tree today. And, indeed, by studying trees rings and the analogous annual growth rings of coral going back 1,000 years and more, the $^{13}C/^{12}C$ ratio is constant to better than 1 part in 10,000. Until, that is, around 1800.

The Industrial Revolution of the late eighteenth and early nineteenth centuries saw the dawn of a major new source of CO_2 for the atmosphere—the combustion of fossil fuels. First coal in the last decades of the eighteenth century, then natural gas in the 1820s and oil in the 1860s were burned to power the explosive growth of manufacturing and steel production that built the modern world. These three fuels are all derived from plant material (with some algae and plankton thrown in) that was alive between 360 million and 66 million years ago (see chapter 12). As with plants today, the lush forests of those ancient C3 plants discriminated against $^{13}CO_2$, meaning that their fossil remains today have low $^{13}C/^{12}C$ ratios. Burning, as we do today, about 15 billion tons per year of these ^{13}C-deficient ancient plants should mean that the total atmospheric ratio of $^{13}C/^{12}C$ will decline.

And that is precisely what we see when we analyze the isotope ratios in old wood. Beginning around 1800, there is a steady decline in the atmosphere's ^{13}C that accelerates as our fossil fuel use grows. The preindustrial level of atmospheric CO_2 was 280 ppm, whereas today it is 420 ppm, a 50 percent increase. Meanwhile, the $^{13}C/^{12}C$ ratio has declined by 0.22 percent between 1750 and today from the release of the ^{13}C-depleted fuel molecules.[8]

How do we know it is the fossil fuels and not, instead, modern plants? They likewise discriminate against ^{13}C, so when they die and/or burn, they also release CO_2 depleted in ^{13}C to the atmosphere. The even heavier ^{14}C isotope provides the answer. Because it is slower moving than $^{13}CO_2$, $^{14}CO_2$ is even less likely to end up in a plant's molecules. But as we saw in chapter 8, enough of the radioactive isotope is incorporated to allow us to use it in determining fluctuations in its formation rate and thus to calibrate C-14 ages. However, the youngest fossil fuels are 66 million years old. This means that 66,000,000 years/5,730 years = 11,518 half-lives have elapsed: fossil fuels thus have precisely zero ^{14}C atoms left. Because we observe a steep decline of 4 percent per decade in $^{14}CO_2$, the obvious conclusion is that the new CO_2 in the atmosphere comes from a source far poorer in $^{14}CO_2$ than modern plants.

A slight complication unfortunately forces us to draw this conclusion from models rather than directly from observations. As noted in chapter 8, the nuclear weapons tests in the atmosphere between 1950 and 1963 added a huge spike of ^{14}C to the planet, initially all of it in the atmosphere. That artificial excess has been assimilating into plants, the soil, the oceans, and so on, for the past sixty years, so the ^{14}C amount has been steadily declining toward the background level. As this book goes to press in 2023, we are just crossing the level seen in 1950. That 1950 level was already lower than the preindustrial concentration because of fossil fuel consumption prior to that date, so we can calculate that the bomb pulse ^{14}C is down to a few percentage points of its initial value, and by 2030, it will be completely gone; the decline due to fossil fuel combustion will continue at about 3.5 percent per decade.[9]

This large rate of decline cannot be explained by the modest discrimination ^{14}C faces in modern plants; it can only be accounted for by the addition to the atmosphere of CO_2 completely lacking in this radioactive isotope—that is, from the long-dead plants in fossil fuels. Human activity is unambiguously altering the chemical composition of the Earth's atmosphere.

TEMPERATURES SINCE THE LAST ICE AGE

As noted above, the climate system is complex. In order to extend our modest, century-long record of past climatic conditions, we need proxies—stand-ins for the thermometers and hygrometers we didn't have 1,000, 10,000, or 100,000 years ago. One of the best proxies we have that spans the entire history of human civilization is the isotope ratios in the rings of trees.

As outlined in chapter 8, tree rings constitute a perfect calendar. Using living trees and those long dead, we have managed to construct a complete record extending back 13,900 years to near the end of the last Ice Age. The simplest climate proxy is a measurement of the ring width—in general, the wider the ring, the better the growing conditions that year (warmer temperatures, sufficient water—although see below for caveats). Wood density is also often used as a proxy. But querying the atoms of wood in each ring directly provides more quantitative information. By sampling wood from the ring corresponding to the year of interest, we can use the stable isotopes of Oxygen, Carbon, and Hydrogen to read the temperature and precipitation rate, just as though we had a little weather station at the tree's location recording these numbers for our future use.

The $^{18}O/^{16}O$ ratio is the principal quantity used to establish temperature. As with any of the isotope ratios we have discussed (C, N, etc.), we need to establish an arbitrary standard value. In this case, we use what is called the Vienna Mean Standard Ocean Water (VMSOW),[10] which is a sample of ocean water collected at depths of several hundred meters from which all salts and other chemicals have been removed. The value for $^{18}O/^{16}O$ is 2005.2 ppm, or almost exactly 0.20 percent.

As should be obvious by now, a water molecule made of $H_2^{18}O$ will be heavier, and thus move more sluggishly, than one made of $H_2^{16}O$. This means two things: (1) it is harder for the heavier molecule to transition from the liquid to the gaseous state (i.e., to evaporate), and (2) as noted in chapter 10, if it does evaporate and ascend to the clouds where it cools (and thus slows down even more), it finds more easily some companions to stick to and make a water droplet again, a drop that eventually falls as rain or snow. The same thing would be true for a water molecule that includes deuterium, the stable heavy isotope of Hydrogen (^2H), although the effect is smaller because the mass difference is less.[11]

The abundance of ^{18}O, then, is greater the higher the temperature. Many studies have been conducted to reconstruct historical temperatures using the Oxygen isotope data in tree rings. T. J. Porter and collaborators examined a 150-year record (1850–2003) from three white spruce trees growing in the Mackenzie River delta of northern Canada.[12] Using thermometric temperature data from the weather station at nearby Inuvik Airport between 1957 and 2003, they showed there was a tight correlation between the measured temperature and the ^{18}O values, where the two sets of data overlapped. This allowed the researchers to extend the temperature record back to 1850; the total $^{18}O/^{16}O$ ratio varied from +1.7 percent to +2.3 percent compared to VMSOW, and clearly shows a rise between 1970 and 2000, where the measurement on the latter date is higher than any other point in the 150-year record. The authors conclude that the temperature in this Arctic location had risen by 1.0°C between 1950 and 2000, twice the increase recorded over that interval at lower latitudes and consistent with the recently observed trend of greater warming in the Arctic.

The same authors also measured the $^{13}C/^{12}C$ ratio that can be used to assess humidity. Trees breathe in CO_2 through tiny pores on the underside of their leaves called stomata. However, stomata are two-way streets—gases flowing through them move both in and out of the leaves. While photosynthesis is busy using these gateways to take in CO_2 and exhale O_2, H_2O can also leave through this route. This so-called transpiration is essential to the tree because the water loss creates a negative pressure that draws in water from the roots all the way up to the leaves, a process critical for the growing wood layer (and the collection of maple syrup for your pancakes). If the relative humidity is high and rainfall is plentiful, the pores remain open with no harm to the tree. But under dry conditions, the tree would lose too much water through their stomata, so they throttle down their openings. This makes it harder for CO_2 to enter the leaf, meaning the tree must be less picky about which Carbon isotopes it will accept. A recent calibration of this effect using trees from Germany's Black Forest shows a rise in the $^{13}C/^{12}C$ ratio of 0.17 percent for each 10 percent decrease in relative humidity.[13]

Climate reconstructions from tree rings back through history show small but significant fluctuations in mean temperature that have had apparently outsize effects on the course of human history. But tree ring reconstructions based simply on ring width or wood density rather than detailed isotopic analysis have faced some criticism because of the complicated interrelationships between the factors that determine tree growth. For example, it has been shown that ring width does increase with a warming climate up to some optimal temperature, but then it begins to decline as the temperature keeps rising, probably as a consequence of heat stress occasioned by excessive transpiration and a shortage of water. It is also likely that the evolving density of the forest and other extraneous factors bias tree ring width.[14]

In 2007, Craig Loehle undertook a project[15] to examine other temperature proxies to see if they mirrored tree ring temperature records. Using an impressive array of studies that examined everything from pollen spores to deep-sea sediments (an excellent tracer of sea surface temperature from the $^{18}O/^{16}O$ ratios found in the $CaCO_3$ shells of foraminifera), ice core data (see below), cave deposits, and other samples from throughout the world, he found a very good match to the European tree ring record over the past 2,000 years. His final plot is reproduced in figure 11.1.

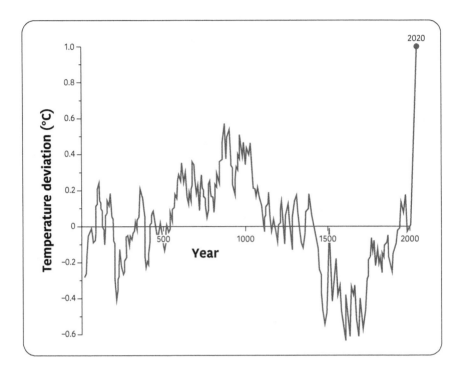

11.1 The temperature anomaly (the difference from the long-term average) in northern Europe over the past 2,000 years derived from tree rings and verified by comparison with other records (ice cores, seashells, cave deposits, etc.) from around the world. The dominant excursions are the well-documented Medieval Warm Period (600–1200 AD) and the Little Ice Age (1450–1800 AD). (after Loehle et al.)

The total excursions over the past 2,000 years range from −0.6°C to +0.6°C (excluding the last decade, during which global temperatures have reached +1.0°C). Two major excursions stand out: the so-called Medieval Warm Period and the Little Ice Age.

Beginning around 600 AD and lasting for more than 600 years, the temperature was consistently above the long-term average, peaking between 900 and 1050. It is no accident, then, that the Icelandic sagas record that Erik the Red led a group of intrepid Vikings to establish a settlement in Greenland in 985 AD (a date corroborated to within a few decades by ^{14}C dating of artifacts found at the site). Fifteen years later, his son Leif Erikson (Erik's son— that's how Icelandic surnames are given to this day) was exploring the North

American coast. Over the ensuing decades, the total population of Vikings in Greenland reached several thousand people along with their cows, sheep, and goats. A Catholic diocese was established in 1126 and ultimately five churches were constructed.

By 1200, however, the temperature had returned to the long-term average, a fall of 0.6°C, where it hovered for two centuries before beginning a rapid plunge in 1400, reaching a minimum 0.5°C below the average by 1450. The last written document found in Greenland was the church record of a marriage in 1408. It is thought that the colony perished (or abandoned the huge island) by 1450. While many theories have been advanced to explain the demise of the colony, the temperature decline was likely a contributing factor. The Vikings needed to harvest enough hay in the summers to feed their animals throughout the harsh winters, and in such a marginal climate, even a change of just one degree can mean the difference between adequate supplies and starvation.

It is also not an accident that, despite the decline in technological sophistication and generalized strife that beset Europe after the fall of the Roman Empire, the period from 1000 to 1300 saw both the building of the great Gothic cathedrals and the mounting of several crusades. A more benign climate leads to great crop yields, decreasing the need for everyone to scratch out life as subsidence farmers, thus producing surplus labor. It is also not coincidental that, as the temperature continued to fall into the sixteenth and seventeenth centuries, large numbers of people from northern Europe undertook the perilous journey to start life again in the New World.

Indeed, the period from 1450 to the early 1800s is referred to as the Little Ice Age. Mean temperatures were 0.5°– 0.6°C below the long-term average and didn't recover completely until 1900. While the coldest period stretched from 1500 to 1700, the year 1816 was also anomalously cold because of a volcanic eruption the year before, the largest in almost 800 years. Tambora, in Indonesia, spread a blanket of reflecting material throughout the Earth's upper atmosphere, thus reflecting sunlight and lowering temperatures by up to 1°C. It snowed in New York on June 6 and New Jersey had a week of frosts later that month. Crops failed in Europe, North America, and China, leading to drastic rises in food prices and famines.

The Little Ice Age, as well as the temperature minimum in the early 1800s, coincided with periods of unusually low sunspot activity. Indeed, within a couple of decades of Galileo's discovery of dark spots on the surface of the Sun, the spots largely disappeared for nearly a century. As discussed in chapter 8, a less active Sun provides less protection against cosmic rays impacting the atmosphere, which in turn influences radioactive isotope production. In particular, the relatively long-lived isotope (compared to ^{14}C) Beryllium-10 (^{10}Be) mentioned in chapter 8 can be measured in both ice and deep-sea cores. It shows peaks up to three times the average value in both the seventeenth and the first decades of the nineteenth centuries, when sunspots were at their lowest level of the last 300 years.[16]

While we know the Sun is about 0.1 percent fainter when sunspots are at a minimum in the eleven-year solar cycle, that small change alone is far from sufficient to trigger global cooling. However, the climate system is complex, and feedback loops can amplify an otherwise small change. For example, the ultraviolet light from an active Sun can be as much as 2 to 3 percent higher (compared to a total irradiance change of only 0.1 percent), and the extra upper-atmosphere ozone (O_3) produced by this enhanced ultraviolet (UV) light could trigger major shifts in atmospheric circulation patterns that have an impact on climate.

Trees around the world have also been helping us construct the climate's impact on the history of nonliterate peoples. For example, in the American Southwest, wood isotope ratios have allowed us to understand the abandonment of the Mesa Verde cliff dwellings. The Mesa Verde site was occupied periodically beginning as much as 10,000 years ago. But beginning around 1150 AD, a massive construction project was undertaken to build the city whose ruins we see today—it housed more than 20,000 people. As the tree ring record shows, a prolonged drought set in from 1276 to 1289 AD, and near the end of this period, the city was abandoned. One hundred miles south, the extensive city in Chaco Canyon, with its five-story, 600-room apartment buildings and massive ceremonial kivas accommodating hundreds of people, was abandoned more than a century earlier in response to the worst drought of the past 1,200 years, which lasted for several decades centered around 1146–1155 AD.[17] Note the coincidence of these culture-defining drought events with the Medieval Warm Period discussed above.

As a final example of deriving climate records from trees, consider an interesting temperature reconstruction from northern Finland extending back to 138 BC. It shows a fairly convincing average decline of 0.31°C per thousand years until, of course, the recent human-induced rapid warming, which managed to overcompensate for that trend in just four decades.[18] As we will discuss below, the long-term climate record derived from Greenland and Antarctic ice suggests we are due for a dramatic cooling over the next dozen millennia or so. The fact that we are instead experiencing rapid warming at an unprecedented rate speaks to the magnitude of humans' impact on the planet.

From its vital role in the calibration of the ^{14}C production rate that greatly improves the accuracy of Carbon dating, to providing records of past climate directly, dendrochronology has become an essential tool in reconstructing the history of the world throughout the period over which human civilization has developed.

LONG-TERM CLIMATE INFLUENCERS

As the data above unambiguously show, human activity is significantly altering the temperature of the planet. Yet we know that much larger temperature fluctuations have occurred in the past; over the last billion years we've seen both a snowball Earth nearly completely covered in ice and a fully tropical Earth with palm trees growing in Greenland. Clearly then, natural forces can also affect the energy balance of the planet. We begin by exploring the dominant causes of temperature change over the past million years prior to human intervention.

The Earth's motion through space is dominated by the gravitational influence of the Sun. Our planet traverses its slightly elliptical orbit once every 365.24255 days. But the Earth and the Sun are not the only two bodies in the solar system, and the Earth feels mutual pulls and pushes from the Moon, Venus, Jupiter, and the other planets. As a result, it experiences subtle and roughly periodic changes in the shape of its orbit, the tilt of its rotation axis with respect to the plane of this orbit, and the orientation of its axis in space (see figure 11.2). The magnitude of these effects and their relationship to Earth's climate were first calculated by the Serbian engineer, mathematician, and astronomer Milutin Milankovic.

The Earth spins once a day around a rotation axis currently pointed at the star Polaris (the North Star), but this orientation is fleeting. Over the course of roughly 23,000 years, the axis traces out a circle in the sky in the same way a spinning top's axis wobbles before it falls over, a dynamical effect called precession. This would be unremarkable (unless you are a celestial navigator) were it not for the fact that the Earth's orbit is elliptical, meaning that over the course of the year, the planet gets closer to and then farther from the Sun. The seasons are determined by which hemisphere is tilted toward the Sun (see figure 11.2a), so

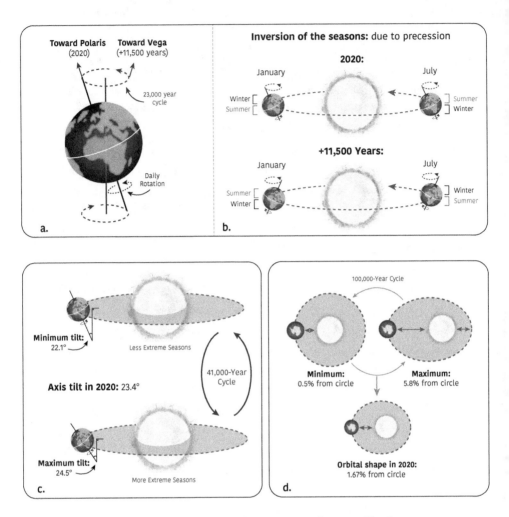

11.2 (*a*) The precession of the equinoxes. The spinning Earth is tugged by the gravitational attraction of the Moon, Sun, Jupiter, and the other planets so that its spin axis slowly changes the direction in which it points, just as a spinning top wobbles in response to the Earth's gravitational pull. The net effect of this precession is that the location of summer and winter in the Earth's orbit reverses on a timescale of approximately 11,000 years, completing one full cycle between 19,000 and 23,000 years. (*b*) Because the Earth's orbit is elliptical, the precession either amplifies or suppresses the difference in seasonal sunlight received (see the text for details). (*c*) The other two orbital effects that constitute the Milankovic cycles. The amount of tilt of the Earth's orbital axis changes from 22.1 to 24.5 degrees and back again on a timescale of 41,000 years. More extreme tilts lead to more extreme seasonal variations. (*d*) In addition, the shape of the Earth's orbit becomes more (up to 5.8 percent from a circle) and less (down to 0.5 percent out of round) elliptical on a timescale of approximately 100,000 years. A more highly elliptical orbit exaggerates the variation in sunlight received over the seasons (see the text for details).

if the Earth is closer to the Sun when the Southern Hemisphere is tilted toward the Sun (as it is today), southern latitudes receive a little more energy than does the Northern Hemisphere six months later when *it* is tilted toward the Sun but is slightly farther away. Given the precession period, this situation will reverse in a little more than 11,000 years, when the Northern Hemisphere is both tilted toward and closer to the Sun during its summer.

In addition to the change in axis *direction* caused by precession, the *magnitude* of the Earth's axis tilt changes slowly over time. It is currently 23.44° away from being perpendicular to the Earth-Sun orbital plane. This will slowly decrease to a minimum of 22.1° and then gradually grow again, reaching a maximum of 24.5° before returning to its current value. The intensity of the seasons depends on the amount of tilt, so this process can also affect the long-term climate. One cycle, from minimum tilt to maximum and back again, takes 41,000 years.

Finally, the shape of the orbit itself changes. Today, our elliptical orbit is 1.67 percent away from being a perfect circle. Even this slight distortion is decreasing and will reach a minimum of only 0.5 percent out of round before increasing again to a maximum value of 5.8 percent. At this upper extreme, the difference in the amount of solar energy striking the Earth between its closest approach and its greatest distance six months later is much more significant than it is today (12 percent versus 3 percent). As we will see, this has a dramatic effect on the Earth's temperature. The time it takes to go from maximum to minimum and back again is roughly 100,000 years.

TEMPERATURES AND THE ATMOSPHERE OVER THE PAST 1 MILLION YEARS

Given these cycles measured in tens to hundreds of thousand years, it is clear we need climate proxies with a much longer memory than the trees if we wish to understand the long-term climatic history of Earth. Happily, such proxies exist in the 3-kilometer-thick layer of ice that covers Greenland and the even deeper layer blanketing the Antarctic continent. These glaciers have been laid down by the gradual accumulation of snow, year after year, millennium after millennium. This snow carries the isotopic signature of the ocean water from which it originated; thus, it provides a direct measure of global temperature extending back more than 1 million years.

Even better, tiny air bubbles frozen in the ice provide us with samples of Earth's atmosphere at the time of their creation—as though we had a foresightful chemist collecting ampules of gas each year over a thousand millennia and leaving them, labeled by date, for our examination. In addition, the ice contains a record of the radioactive isotopes produced by cosmic rays, allowing us to assess the activity of the Sun for a hundred times longer than the tree ring record. Thin deposits of volcanic dust from distant eruptions allows us to extend the record of our planet's geologic history, while dirt from as far away as the Gobi Desert, found in the Greenland ice, indicates the waxing and waning of stormy conditions. It has even been suggested that isotopes from nearby exploding stars might be present in the ice record.

To read this dramatic and extensive history, we just drill a vertical column through the ice, extracting a core about 10 cm (4 inches) in diameter and from one to a few meters (3 to 15 feet) in length. The cores are catalogued and often stored temporarily in a snow cave dug near the drilling site. Eventually they are transferred to a permanent storage location such as the National Science Foundation Ice Core Facility in Lakewood, Colorado, where the 17 km (more than 10 miles) of ice are kept in a 55,000 cubic foot room at $-36°C$ ($-33°F$). Scientists from around the world come to extract samples from the cores to study the many secrets they hold.

One important parameter derived from the ice is the long-term temperature history. As with trees, the Oxygen and Hydrogen heavy-to-light isotope ratios are used. Our VMSOW defines the starting value in the ocean from which the water evaporates to make clouds. As noted above, the heavier ^{18}O isotope has a little more trouble breaking free from its liquid neighbors, and so the clouds that form have an $^{18}O/^{16}O$ ratio about 0.8 to 1.0 percent lower than the water from whence it came. As the cloud drifts northward or southward and cools, its water vapor recondenses into drops of rain. The heavier, slower-moving water molecules containing ^{18}O condense more readily and fall out sooner, leaving the clouds even more depleted in the heavy isotope. By the time they reach Greenland or Antarctica, the snow that falls there may be as much as 5.0 percent lower in ^{18}O content.

The exact values, of course, depend on the temperature—the higher the temperature, the faster the $H_2^{18}O$ molecules jiggle and the easier it is for them to escape into the air. For every 1.5°C increase in temperature, the ^{18}O content increases by about 1 ppm or 0.05 percent of its usual value. Such a change is easily measurable, so we can reconstruct the average temperature to a fraction of a degree over the entire length of the core, back to 800,000 years for the Dome C core from the European Project for Ice Coring in Antarctica (EPICA).[19] The ratio in recently formed ice is 3.4 percent below the standard, while at the peak of the last Ice Age 25,000 to 30,000 years ago, the ratio was 4.6 percent low.[20] This corresponds to a global temperature change of about 10°C (18°F). The values from Greenland and Antarctica, while arising from different parts of the ocean, show excellent agreement; the largest discrepancy is 0.05 percent, and a more typical difference is less than 0.015 percent (less than 0.4°C).

Another vitally important datum measured in the ice cores is the atmospheric composition as derived from the trapped bubbles of air. A single bubble 1 millimeter in diameter (0.04 inches across) will contain about 10,000 trillion molecules of air, so even trace constituents (e.g., N_2O at 0.3 ppm) will be represented by billions of molecules. Thus, highly accurate composition measurements are possible. The results are striking. For example, methane, a potent greenhouse gas, rose and fell every 100,000 years in synchronization with the Earth's orbital shape change, varying from 400 parts per billion to 600 parts per billion. But around 1820, it began a dramatic rise and today has a concentration of 1,920 parts per billion, a 380 percent increase.[21] N_2O shows a similar pattern, although with a more modest rise of 30 percent.

Most interesting of all is the record of CO_2. Over the past 450,000 years, it has shown a minimum concentration at the peak of the Ice Ages of 180 ppm (constant over the past four 100,000-year cycles to within 2 percent). After hitting this minimum, it shows a rapid (in geologic terms) rise to 280 ppm. Then, within a few thousand years, it begins to decline toward the next minimum, with some bumps and wiggles along the way. The remarkable fact, however, is that, if we plot the temperature from the ^{18}O record and the CO_2 concentration on the same graph, they track each other almost perfectly (see figure 11.3).[22] The fluctuations exhibit signatures of all three orbital cycles—the precession at 23,000 years, the axis tilt change at 41,000 years, and the dominant 100,000-year orbital shape change.

The only exception we see to this tight correlation is for the present epoch; CO_2 today is at 420 ppm, a 50 percent increase from the level seen in preindustrial interglacial periods, yet the temperature is only 1 degree higher. We know already where all those extra molecules came from, but why don't we see a 10°C temperature increase, as the graph in figure 11.3 would predict? The temperature and CO_2 have tracked each other so closely in the past, strongly suggesting a

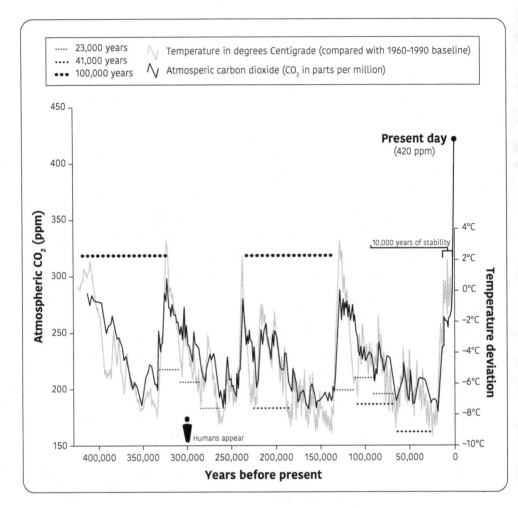

11.3 History of Earth's climate over the past 450,000 years as derived from Antarctic ice. The left axis and black line show the variation in CO_2 concentration in the atmosphere. The CO_2 concentration varies cyclically from 180 ppm to 280 ppm, with a roughly 100,000-year period (thick black dashes). The right axis and gray line show the mean temperature as derived from the $^{18}O/^{16}O$ ratio. The degree to which the temperature and CO_2 values track each other is remarkable. Also indicated in the figure are periods corresponding to the other two Milankovic cycles (varying axis tilt at 41,000 years and precession at 23,000 years), both of which also show up in the data. Note that the last 10,000 years over which modern civilization developed is the most stable period in the past half million years. The current CO_2 concentration is far higher than it has been in millions of years.

causal connection. Why aren't we already insufferably hot? The answer lies in the enormous capacity of the world's oceans to absorb energy.

The rate at which the Earth receives energy from the Sun, averaged over day and night, all latitudes, and the seasons, is 342 watts per square meter (342 W/m²—a bit less than six 60-watt light bulbs arrayed over each 3-foot by 3-foot patch of the Earth). As noted earlier, 31 percent of this energy is reflected back into space, leaving 236 W/m² to warm the planet. All the greenhouse gases we have added to the atmosphere over the last 250 years have the net effect of adding just over 3 W/m² to this value; that is, the blanketing effect of the extra CO_2, CH_4, and other gases is equivalent to having turned up the energy from the Sun by 1.25 percent. While 3 W/m² may not sound like much, integrated over the surface of the Earth and summed over a year, it means a total net energy increase of 50 billion trillion joules (5×10^{22} J). Since it takes only 0.004 percent that much energy to raise the temperature of the atmosphere by 1 degree, why isn't it already warmer?

Water has four times the heat capacity of air; that is, it takes four times as much energy to raise a kg of water 1 degree as it does to raise a kg of air by the same amount. And the oceans contain *many* more molecules than the atmosphere—just about 1 million times more. Thus, the Earth's water has the capacity to absorb 4 million times as much energy as the air. It can't do that instantaneously, of course, because only the upper layers of the ocean are directly exposed to the extra heat the molecular blankets are producing. But it is estimated that more than 93 percent of the extra heat the Earth has received over the past fifty years has been absorbed by the ocean.[23] Indeed, the sea surface temperatures reflect this, having risen an average of 1.5°C since 1900.[24] Water's huge capacity to absorb energy explains why the air temperature has risen only a "modest" 1°C.

The warming of the ocean is not without consequences, however. First, with all those water molecules jiggling faster, they take up more space. Fully half of the 10 inches of sea-level rise we have witnessed over the last century is due to the thermal expansion of the warming water. Warming water also speeds the melting of ocean ice, decreasing the reflectivity of the planet and inducing further warming.

This type of interaction, where the outcome effects the original cause, is called a feedback loop. In this case, the feedback loop is positive—warming waters melt more ice, reducing the reflectivity, increasing the heat absorbed, which leads to more warming of the water and more melting of the ice. The effect amplifies the cause. Despite the connotation of the label "positive," this kind of feedback is usually bad; it leads to a runaway situation (as in the *Sorcerer's Apprentice*). Sometimes the effect suppresses the cause, producing a negative feedback loop: warmer oceans lead to more evaporation, which leads to the formation of more clouds, which reflect more sunlight, so less energy is absorbed and the oceans get cooler. Negative feedback loops are stable—if you push the system in one direction, the result pushes back. One of the principal difficulties in modeling the Earth's climate is that the many parts of the system are linked by dozens of such negative and positive feedback loops

The rise in ocean temperature has numerous other implications. Much of the life in the sea is very sensitive to temperature. Warming oceans have effects ranging from the bleaching (i.e., death) of coral reefs to forcing New England lobsters to migrate to Canadian waters; from 1996 to 2014, the lobster catch in New York and southern New England fell by 97.7 percent.[25] From plankton at the base of the food chain to blue whales, the largest creatures ever to live on Earth, fluctuations in the ocean temperature reset ecological balances, change migration patterns—and, as we shall see in chapter 13, drastically alter the global circulation patterns of water and air, profoundly changing the climate itself.

WEATHER AND CLIMATE

When discussing the impact of humans on our atmosphere, it is important to distinguish between weather and climate. Weather involves changes in local atmospheric conditions on timescales of minutes to years. It involves phenomena such as storms, warm fronts, droughts, and cold snaps. If you are wondering whether you should take an umbrella on your way out the door today or whether you should buy a warmer parka for the coming winter, you are making decisions based on the weather.

Climate involves global changes on timescales of decades to millennia and beyond. The ice ages, Medieval Warm Period, and the current forty-five-year run of ever-increasing temperatures represent changes in the climate. Decisions about buying a beachfront home or farmland in Canada are decisions that should be informed by a knowledge of climate trends (this author is certainly avoiding the former and considering the latter).

Attributing any particular hurricane or snowstorm or seasonal drought directly to climate change is generally unwise. Long-term climate changes certainly bias weather patterns, making extreme heat and rainfall events more probable, for example, but saying, "It's pretty damn hot today—must be because of global warming," is not consistent with careful definitions of weather and climate. By the same token, "global warming" does not mean the end of cold winters and snowstorms—Senator Inhofe's snowball in the Senate chamber notwithstanding.[26] Indeed, colder and stormier winter weather in mid-latitudes is predicted by global climate models as the planet warms. In one recent event, the direct connection between global warming and a winter snowstorm in Europe has been demonstrated using our friendly isotopes.

In late February and early March 2018, Europe was hit by major snowstorms that significantly disrupted life across the continent, reaching as far south as Rome, where the city was blanketed in white for only the second time in thirty-five years. Isotope geochemist Hannah Bailey from the University of Oulu in Finland and her colleagues[27] analyzed the isotopic composition of the snow itself, as well as the water vapor in the air blowing off the Barents Sea north of Finland. Both showed a striking change in the ^2H to ^{18}O ratio of 1.6 percent in a single day (February 19), followed by a rise to 3.1 percent roughly a month later. Because ^{18}O has two extra neutrons and ^2H has only one, $^2H_2^{16}O$ evaporates more easily than $^1H_2^{18}O$ (although both are slower than $^1H_2^{16}O$). This allowed the authors to conclude that at least 88 percent of the snowfall in Europe came from evaporated Barents Sea water, water that was available for evaporation at all only because of the drastic fall in sea-ice cover, itself a direct consequence of global warming. They showed that the monotonic decline in sea-ice cover of the region (60 percent ice loss in the preceding forty years) led directly to a steady increase in maximum March snowfalls across Europe, and they predict that within fifty years, when the Barents Sea is expected to be ice-free, heavy winter snowfalls across Europe will be the norm—yes, Senator Inhofe, in a steadily *warming* world.

VOLCANOES, SOLAR STORMS, AND EXPLODING STARS

The atomic historians of the ice cores allow us to reconstruct Earth's climate history over a million years. They tell a story of natural fluctuations in global temperatures of up to 12°C triggered by changes in the Earth's orbital parameters and amplified by the compositional changes in the atmosphere, along with the many feedback loops that accompany these changes. They also reveal the unprecedented, rapid impact of the vast geochemical experiment human activity is now inflicting on the atmosphere and allow us to test our models for the effects this experiment will have on the planet's future.

Snow is not the only thing that falls on the ice caps. Dust whipped up by desert storms can remain airborne for weeks and travel thousands of miles before "raining out" of the atmosphere (e.g., Saharan dust not infrequently reaches Florida). Similarly, salt from ocean spray becomes airborne in storms and settles on the ice. Explosive volcanoes can eject a quarter of a trillion tons or more of gases, rocks, ash, and glassy shards into the stratosphere 20 miles above the surface, where it drifts all the way around the globe and takes a few years to fall out. Volcanoes in particular often emit a lot of sulfurous gases (primarily SO_2) that, upon reaching the stratosphere, react with water to form H_2SO_4 (sulfuric acid). This in turn condenses into shiny droplets that reflect a lot of sunlight back into space before it can warm the Earth. Eventually, the material settles back to the ground, leaving a layer of sulfur-containing compounds on the ice. (The sulfuric acid is oxidized under the influence of the UV light that is plentiful above the Ozone layer in the lower stratosphere and becomes sulfate—SO_4.) Finally, radioactive isotopes produced by cosmic rays such as ^{10}Be and ^{14}C—and even rarer isotopes produced by exploding stars—settle on the ice and have stories to tell.

An instructive example of using multiple proxy records to refine our historical reconstructions is the work by M. Sigl and collaborators.[28] They used tree ring estimates of summer temperatures from ^{18}O, sulfate (SO_4) tracers of volcanic eruptions in ice cores from both poles, and a comparison of the ^{10}Be isotope concentrations in the ice and ^{14}C concentrations in the trees. They found that, since 1250 AD, the trees showed that the temperature cooled one to two years after large volcanic eruptions (as we measured directly following the Mount Pinatubo eruption in 1991). But for the preceding 2,000 years or so, there was a seven-year lag between the date of an eruption (as measured in the ice) and the Earth's cooling (as measured in the trees). In addition, there was a seven-year offset between two ^{10}Be spikes in 775 AD and 987 AD and a similar jump in the ^{14}C from the tree ring record (seen around the world in 782 and 994 AD, respectively). This represented prima facie evidence that there was a seven-year error in the ice chronology starting about 800 years ago.

The new dating sequence allowed an uncertainty in the ice core record of just a year or two all the way back to 2,500 years ago. Over this period, they chronicled 283 individual volcanic events large enough to leave signatures in Greenland and/or Antarctica; the rough latitude of the eruptions can be deduced from the relative strength of the sulfate signals at the two poles. While only eighty-one of these (29 percent) are classified as likely located in the tropics, those eighty-one account for nearly two thirds of the total climate impact over the last 2,500 years. Five of these eruptions—in 426 BC, 44 BC, 536 AD, 1257 AD, and 1458 AD—produced a greater sulfate signal than the 1815 Indonesian eruption of Tambora discussed earlier that led to the year-without-a-summer; the 426 BC and 1257 AD events producing nearly twice the effect of Tambora, implying the ejection of roughly 40 cubic km of material.

The 1257 event has recently been identified with the Samalas volcano, also in Indonesia. Records written in Old Javanese on palm fronds recount a devastating eruption sometime before the end of the thirteenth century. Wood found buried at the base of the debris on Lombok Island, roughly 100 miles east of Bali, where the volcano is located, has ^{14}C dates consistent with a 1257 eruption. Furthermore, glass shards of the type produced in explosive volcanic eruptions are found both on the island and embedded in the ice caps 10,000 km away. They have exactly the same abundances of 69 to 70 percent SiO_2 and 8.0 to 8.5 percent Na_2O+K_2O, making a common origin assured. It is clear this eruption was a worldwide event, and the impact on the climate was substantial.[29]

Cooling estimates for central and northern Europe during the summer of 1258 amount to 1°C (and remember what happened to the Vikings with that kind of temperature decline). Written records from the time describe 1258 as follows:

What, then, shall I say about the fruits of the earth that year, when the weather was so remarkably unseasonable that the warmth of the Sun was hardly able, even a little, to reach the earth, and the fruits of that year could barely attain maturity, if at all? For so great a thickness of clouds covered the sky throughout that whole summer that hardly anyone could tell whether it was summer or autumn. The hay, drenched incessantly by strong rains that year, was unable to dry out, because it could not collect the warmth of the Sun on account of the thickness of the clouds.

—Richard of Sens, writing in 1267

A well-documented account of the social upheaval the weather caused includes food price hyperinflation, famine, sickness among both farm animals and humans throughout Europe and the Middle East, and a full lunar eclipse (only possible when a thick stratospheric layer of dust is in place) has been given by Richard B. Stothers.[30]

The ice core volcanic record can be extended much further back in time. A. V. Kurbatov and colleagues have analyzed an Antarctic record extending back 12,000 years; they found seventy-two large eruptions between 500 BC and 10,000 BC. The largest single eruption in this period (as measured by the concentration of SO_4 in the ice) occurred 7,881 years ago; needless to say, no written records are available from that time.[31]

Deserving of further mention is the coincidence of the ^{10}Be and ^{14}C spikes from 775 AD and 987 AD found in the ice cores and tree rings, respectively, and used in the Sigl study[32] to calibrate the calendar ages of the ice cores. What induced the sudden increase in cosmic rays that boosted the production of these two radioactive isotopes so quickly and dramatically? A subsequent study of the ice record by F. Mikhaldi and his collaborators added another isotope, Chlorine-36 (^{36}Cl), to the story and provided strong evidence for an answer.[33] Like ^{14}C, ^{36}Cl exists in a ratio of about 1 in a trillion compared to the element's two stable isotopes, ^{35}Cl and ^{37}Cl (discussed in chapter 5); it is also produced in the atmosphere by cosmic rays.

The researchers found that the relative amounts of the three isotopes were consistent with what should be expected from solar cosmic rays. They conclude, however, that the size of the solar flare required to produce these abundance spikes had to be substantially greater than the largest solar storm on record, the Carrington event of September 1 and 2, 1859.[34] That event produced global effects, with the aurora borealis seen as far south as the Caribbean, and bright enough in New England to read a newspaper outdoors at midnight.[35] Almost all of the world's telegraph systems failed because of the huge currents induced in the Earth's magnetic field. If an event even this large, let alone several times greater, were to occur in today's satellite-dependent, electronically connected world, it would be catastrophic. Lloyds of London estimated the cost to the U.S. economy alone to be between $0.6 and $2.6 trillion.[36] This is a case in which we very much hope the past is not a preview of the future.

Finally, we briefly mention the matter of even more distant events affecting the ice core record. In a matter of hours, the explosive destruction of a star releases more energy than the Sun will produce over its entire 10-billion-year lifetime (see chapter 16). Thus, even though the closest such events recorded over the last 2,000 years are hundreds of millions of times farther away than the Sun, their enormous energies could still affect Earth. In particular, a burst of X-rays impinging on the upper atmosphere would ionize a lot of the Nitrogen to N_2^+, which in turn would combine with Oxygen in a series of steps to make nitrate ions (NO_3^-). Various claims have been made over the past forty years that spikes of NO_3^- have been found in the ice that correspond to stellar explosions observed in the years 1006, 1054, 1572, and 1604 AD, but none has been convincingly replicated in an independent ice core.

With only a few stars exploding every century in our entire galaxy, one should expect to wait a long time—more than the million years of the ice core record—before a nearby event really lights up the sky. Recently, however, three researchers at the National Institutes of Health connected an increase in the rare isotope Iron-60 (^{60}Fe) in a deep-sea core with our passage near a cluster of stars likely to have produced several explosions about 2.5 to 3 million years ago. The nearest such explosion would have been at least thirty times closer than those we know about in the historical record, making the impact on Earth 900 times as great.[37] Because ^{60}Fe is radioactive, with a half-life of 2.6 million years, roughly half of any that we received from those explosions would still be present today. There is no source of this isotope on Earth; all ^{60}Fe currently present arrives as a tiny, tiny fraction of the 10,000 tons or so of interplanetary and interstellar dust that drifts down on Earth each year.[38]

Indeed, through a Herculean effort, ^{60}Fe has recently been discovered in Antarctic snow. A team from Germany collected 1,100 pounds of snow from their Antarctic research station (all of it less than twenty years old) and transported it in the frozen state to Munich. There they melted it, carefully filtered it, and passed the nonwater components through a mass spectrometer. They found five atoms of ^{60}Fe (out of the 50,000 trillion trillion atoms making up the original snow sample). This establishes the rate of ^{60}Fe accretion as about three atoms per square foot each day, or less than 0.6 milligrams (the mass of two poppy seeds), over the entire Earth in a year. Through careful comparison of the amount of ^{60}Fe with other isotopes, the authors systematically ruled out origins from solar system meteoritic material, nuclear bomb tests, and other human effects, concluding that these atoms come from the interstellar cloud of gas and dust through which the solar system is now passing. They note that constructing a long-term record of the ^{60}Fe concentration would provide a record of the regions of space through which the solar system has passed as it orbits the center of the galaxy.[39]

OTHER CLIMATE PROXIES

Tree rings and ice cores are far from the only sources of the climatological record extending into the distant past. Coral reefs have annual growth rings and are also what they eat, of course, and so leave isotopic signatures of the ocean temperature, the cosmic ray–induced elements, and the chemistry of the water in which they grow. Foraminifera,[40] single-celled organisms that encase themselves in tiny calcium carbonate ($CaCO_3$) shells, come in a huge variety of shapes and sizes, and their population mix is acutely sensitive to the temperature and acidity of the ocean. As they die and fall to the bottom, they accumulate in layers, just like the ice, and cores drilled and extracted from the ocean floor provide a continuous record of sea surface temperatures and ocean chemistry. Worms and other creatures that live in the deep ocean mess with the annual layers a bit, but ^{14}C dating gives accurate dates back at least 50,000 years. The $^{18}O/^{16}O$ ratio in the shells follows the reverse pattern of that ratio in the ice beacuse the colder the temperature, the less of the heavy isotope gets evaporated and the more is thus left behind in the ocean, so cold means high ^{18}O in the shells and low ^{18}O in the ice.[41]

Ocean cores can reveal a historical record more than 100 times longer than the deepest ice core. Recently, scientists from Japan retrieved a core sample 75 meters long from 3.5 miles deep in the ocean that contained material from about 100 million years ago. They found many species of bacteria in the sediments and, after feeding them lunch, saw them start to grow and divide—not exactly Jurassic Park but living creatures from the same era brought to life in our own time.[42]

We can extend the climate record, albeit much more crudely, back to the origin of life on Earth 3.9 billion years ago (see chapter 13). There have been many ice ages, or, perhaps more appropriately, ice eras, in Earth's history punctuated by long periods where there was no ice on the planet at all. Antarctica became glaciated beginning about 34 million years ago, although extensive glaciers in the Northern Hemisphere appeared only 3 million years ago. Prior to the formation of the current Antarctic ice sheet, the Earth had been ice-free for over 200 million years. One of the most severe glaciations occurred between about 720 million and 635 million years ago, when the Earth turned into a giant snowball, almost completely covered by ice. This was followed by a warm period lasting almost 200 million years that included the Cambrian explosion (540 million years ago [Mya]), when multicellular life became common and migrated onto land for the first time. These long-lasting swings in climate were driven primarily by plate tectonics, the motions of the continents colliding and breaking up again as they float around on the viscous fluid of Earth's mantle.

As described above, the Earth's orbital cycles act slowly on timescales of tens to hundreds of thousands of years. To fully understand the many factors that control our planet's climate—and thus to build confidence in our ability to predict its future—we need climate records even longer than these cycles. By interrogating our atomic historians found in tree rings, ice cores, coral reefs, and ocean sediments, we now have the records we need extending back in time long before the first appearance of *homo sapiens*. Our single species has emerged as a force of nature itself; its collective activities are transforming our climate with unprecedented rapidity. The paleoclimate records are crucial in allowing this species to assess and perhaps even mitigate its impact and thus make its own future less uncertain.

CHAPTER 12

The Death of the Dinosaurs

An Atomic View

As highlighted in chapter 11, it is not the size of the current changes in Earth's temperature, the extent of its ice sheets, or the chemical composition of its atmosphere that are unprecedented. It is the rapidity with which these changes are taking place—over decades rather than tens of millennia—that is both stunning and worrying. Even this rapid change is not unique in Earth's history, however. The last time it occurred, it wasn't simply worrying to the creatures that dominated the planet at the time—it heralded their complete demise.

Sixty-six million years ago (Mya), an asteroid 10 km across slammed into a spot just off the coast of the Yucatan Peninsula in Mexico. The fallout from this event wiped out 75 percent of all species on land and in the oceans around the entire planet, from the diatoms to the dinosaurs. As always, we can use our atomic historians to date the event precisely and to reveal its aftermath from minutes following the collision to effects felt only millennia later.

THE HISTORY OF LIFE ON EARTH

In 1860, a year after the publication of Charles Darwin's *On the Origin of Species*, John Phillips, Oxford professor, president of the Geological Society of London, and well known for positing the first geologic timeline correlating fossils and rock layers, published a book entitled *Life on Earth: Its Origins and Successions*. From his examination of marine fossils, he reported a boundary between the Mesozoic and Cenozoic eras in the geologic record between what we would today call the Cretaceous and Paleogene periods. He showed that several genera,[1] while plentiful up to the end of the Mesozoic era, completely disappeared from the fossil record thereafter.[2]

Over the ensuing century, rapid expansion in the number and variety of fossils collected around the world led to the recognition that it was not uncommon for whole families of related species alive in one era to disappear suddenly in the next. These so-called extinction events have punctuated the history of life ever since the blossoming of multicellular organisms in the Cambrian explosion 540 Mya. While the exact number of such events remains a matter of debate, it is generally agreed that five qualify as "mass extinctions," points in life's history when at least half of all extant species vanish in a geological instant (i.e., less than a couple of million years). These cataclysmic events occurred at roughly 445 Mya, 370 Mya, 250 Mya, 201 Mya, and 66 Mya.

The potential causes of mass extinctions are numerous and much debated. Candidates range from documented geologic events on Earth—flood basalts (see below), global cooling and warming, drastic sea-level changes caused by ice ages or plate tectonics, and geomagnetic reversals (chapter 8)—to astronomical causes such as asteroid or comet impacts and nearby stellar explosions. A third set of options involves the overpopulation and overconsumption of a single species; it has been argued that we are currently in a sixth such mass extinction triggered by one such species—us.[3]

The most recent extinction (barring the present) is the one first identified by Phillips that we have now dated to 66 Mya (see below). This event saw the sudden disappearance of all nonavian dinosaurs after they had dominated the planet for over 200 million years, along with 75 percent of all other plant and animal species. This point in the fossil record is referred to as the Cretaceous-Paleogene (K-Pg) boundary, after the two geologic periods for which it serves as the dividing line.[4] As with other mass extinction events, the cause of this drastic transformation of life on Earth has been much debated. But an investigation begun in the late 1970s by a young geologist at Berkeley led to a dramatic hypothesis that, after forty years of concentrated effort, is now widely accepted. The dinosaurs were killed by an asteroid striking the Earth. Our atomic historians allow a detailed reconstruction of this cosmic collision and its aftermath.

FINDING THE CULPRIT

Osmium, Iridium, and Platinum are elements number 76, 77, and 78, respectively, in the Periodic Table. Being among the heaviest elements with stable isotopes, they are rare in the universe (see chapter 16). They are even rarer on the Earth's surface because, like many of the other heavy elements, they sank to the core of the planet during its molten stage of formation. Indeed, it is thought that a significant portion of these elements now found on Earth come from the rain of pulverized meteoric material—interplanetary dust—that dumps roughly 10,000 tons a year onto the Earth's surface from outer space.

Since the accumulation rate of this solar system dust is thought to be very steady, the abundance of these rare metals can be used as a measure of the sedimentation rate in ocean core sediments and in rock layers formerly under the sea that have since been thrust up onto land. These elements act like a steadily ticking clock. If the sedimentation rate is high and a lot of detritus sinks to the bottom of the ocean in a short time, these heavy elements will be sparsely distributed throughout that layer of rock because they haven't had much time to accumulate. When the sedimentation rate is low, they will be present at a higher concentration because less oceanic debris will have accumulated in the same interval of time. Walter Alvarez was interested in using this phenomenon as a clock to study the time around the K-Pg boundary.

In the Apennine Mountains about 100 miles north of Rome near the town of Gubbio, there is an outcrop of layered limestone that used to be on the floor of the ocean but was pushed up onto land when the northward moving African continent collided with the main European landmass. These rocks contain a record of seafloor accumulation extending from 185 to 30 Mya. Close examination of the layers near the K-Pg divide shows that both the foraminifera (single-celled plankton encased in tiny shells) and their plantlike colleagues, the coccoliths, undergo complete population transformations within a centimeter or two of the boundary layer; virtually all species found below the boundary disappear abruptly, only to be replaced by another set of species above this line. The boundary itself is marked by a 1-cm-thick layer of clay. Alvarez was curious about the sedimentation rate (known to vary from 0.01 to 3 millimeters per century) in the clay layer; was the rate at this sharp boundary faster or slower than in the adjacent rock layers?

He extracted rock samples from just above and just below the boundary layer as well as seven samples from throughout the long (325-meter) Cretaceous record and two extractions from the clay layer itself. The samples were subject to neutron activation analysis—yes, just like the paintings in chapter 7: irradiating the samples with neutrons to induce radioactive isotope creation and measuring the gamma rays emitted. The energy of the gamma rays tied the isotopes to specific elements. A total of twenty-eight different elements were examined— from common ones like Aluminum and Iron to rare atoms such as Lutenium and Tantalum. Although twenty-seven of the elements examined showed slowly varying, parallel trends throughout the long history of the rock, including in the boundary layer, one element did not.

Iridium was the outlier. The concentration in most layers throughout the rock samples was about 0.3 parts per billion (ppb), comparable to what other workers had found in other sedimentary rocks. This simply reflects the steady rain of Iridium-rich material from space. But in the boundary layer of clay, the value was 9.1 ppb, an increase by a factor of 30 (3,000 percent!). None of the other twenty-seven elements changed by even a factor of 2. By 10 cm (4 inches) above the boundary layer, the Iridium level had fallen by a factor of 10, and 1 meter above the boundary, the levels were back to normal.

To test whether this anomaly was the result of some weird local process, Alvarez and his team analyzed the same K-Pg boundary layer from an outcropping in Denmark, 1,000 miles to the north. These sedimentary layers were laid down in significantly shallower waters, so they differed somewhat in composition from the Gubbio rocks, but the layer of clay at the boundary was again clear. In this case, the background level of Iridium away from the boundary was measured as 0.26 ppb, while at the boundary itself, the concentration jumped by 16,000 percent! It was like expecting a small glass of water and getting a ten-gallon bucketful instead. In his paper,[5] Alvarez also cites an unpublished result that a sample from New Zealand, half the world away, shows a similar dramatic Iridium increase right at the K-Pg boundary.

After considering all possible sources of this striking anomaly, only one conclusion made sense: rather than a sudden change in the sedimentation rate, there must have been a sudden change in the accretion rate. That is, all that extra Iridium came from space. Considering the worldwide distribution of the element in this single layer and the Iridium concentration measured in meteors that have fallen to Earth (about 500 ppb), they concluded the incoming culprit must have been about 7 km across (4 miles wide) and must have had a mass of about 300 billion tons. Knowing the typical relative velocity between the Earth and any approaching asteroid, it is easy to calculate the kinetic energy of such an event: roughly 100 billion trillion joules, or 10 to 20 billion times the energy of the atomic bomb that destroyed Hiroshima at the end of the World War II. Such an event would certainly have worldwide consequences, and the authors went on to spin a plausible tale of how, within weeks, the impact killed a large fraction of all living things on Earth.

This hypothesis for the death of the dinosaurs was controversial. Indeed, the first author of the paper reporting these results was not Walter Alvarez, but his father, Luis Alvarez, who just happened to have a Nobel Prize in Physics (and did a lot of the Iridium identification work). One wonders how easy the paper would have been to publish without the first author's reputation. In fact, I recall vividly the colloquium describing this hypothesis that Luis Alvarez gave in the fall of 1980 in the physics department at Columbia University. Our department's Nobel laureates (all three of them) were, as usual, seated in the first row. Questions flew fast and furious. At the end of the talk, many of my colleagues walked out shaking their heads as if to say, "Isn't it sad how great minds decay with age" (Luis was only seventy at the time.) I, on the other hand, thought the meteor hypothesis sounded great!

Over the last forty years, the ever-growing body of evidence supporting this hypothesis has become overwhelming. Its principal competitor—the massive volcanic eruptions occurring over millions of years that produced the flood basalts called the Deccan Traps on India's western edge—arguably did have a long-term impact on the climate given the large release of CO_2 expected to accompany such lava flows, but it is clear that the coup de grace for the dinosaurs was the asteroid.

SITE OF THE COLLISION

When Alvarez and his colleagues proposed the impact hypothesis, one big piece of evidence was absent: the crater. An approximately 10-km asteroid striking the Earth with the expected impact velocity should leave a crater at least 10 to 20 miles deep and more than 100 miles across. Bodies in the solar system without atmospheres and liquid water such as the Moon, Mercury, Mars, and the moons of the outer planets show many craters this size and larger. But the relentless erosional forces of wind and water—not to mention the slowly shifting tectonic plates—means craters on Earth's surface are not apparent for very long (geologically speaking). The famous meteor crater east of Flagstaff, Arizona, less than a mile across and 600 feet deep, is thought to have been created by a space rock about 100 feet across. It is only about 50,000 years old. With more than 1,000 times longer than that for erosion and continental drift to play their parts, it was thought that even the much larger crater expected from the killer asteroid was unlikely to be found.

Two years before Walter Alvarez's discovery of the Iridium anomaly, however, a couple of engineers working for the Mexican national oil company Pemex matched a magnetic field anomaly map they had made by flying over the Yucatan Peninsula of Mexico with gravity anomaly data[6] collected decades earlier. They found that both data sets revealed a perfectly circular pattern, a 180-km-diameter ring structure centered just off the northern coast of Yucatan. Because the survey data were proprietary, this result was not published in the scientific literature. It took another decade before accumulating evidence from around the Caribbean and cores drilled into the Yucatan rocks revealed the Chicxulub[7] crater as the site of the fateful event.

The "where" had been found, but the "when" and the "what" (happened) remained to be established. Four decades and hundreds of scientific studies conducted at locations around the world have now reconstructed that fateful day with remarkable precision.

DATING THE EVENT

Several methods have been used to establish the date of the impact. One of the oldest geologic dating methods involves comparing a pair of ratios for Uranium and Lead isotopes. Other techniques require measurements of Potassium and Argon, and Rubidium and Strontium. In each case, a long-lived radioactive isotope decays to a stable daughter—the so-called radiogenic nucleus—and the relative number of each in a sample reveals its age.

Zircons are minerals containing element number 40, Zirconium, with the composition $ZrSiO_4$. They are formed at high temperatures resulting from volcanic activity or impact events, they are extremely stable, and they range in size from a fraction of a millimeter to a centimeter or more. An important feature of this mineral's crystallization is that it readily incorporates Uranium atoms into its crystal structure but adamantly discriminates against Lead. The two naturally occurring isotopes of Uranium, ^{238}U and ^{235}U, decay through multistep chains of alpha and beta decays to ^{206}Pb and ^{207}Pb, respectively (both of which are stable).

By measuring the ratios of ^{235}U/^{207}Pb and ^{238}U/^{206}Pb and plotting them as the *x* and *y* axes of a graph, one obtains a so-called Concordia line; because the half-lives of ^{238}U (4.47 billion years) and ^{235}U (710 million years) are known, these ratios provide unique ages. For example, given a sample exactly 1 billion years old and starting with 1.0 gram of each Uranium isotope (much too large an amount to be found in nature but useful for illustrative purposes), we'd expect $(1/2)^{1/4.47}$ of the ^{238}U, or 0.857 g, to still be present along with a buildup of 0.143 g of ^{206}Pb, so the ratio ^{206}Pb/^{238}U would be 0.167. For ^{235}U with its shorter half-life, $(1/2)^{1/0.710}$, or only 0.377 g, would be left, so the expected ratio is 1.65 (see figure 12.1).

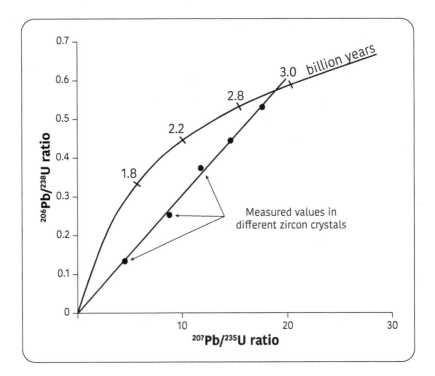

12.1 A Uranium-Lead concordia diagram. The curved line represents the ages for the Pb/U ratios calculated from the respective half-lives of the two isotopes of Uranium. The straight line fits the measured data points for the observed ratios, indicating that varying amounts of the radiogenic Lead have been lost from the measured samples. The intersection of the two lines marks the point of no Lead loss and thus the actual age.

If any Lead has leaked out of the sample over time, the ratios will not fall, as expected, on the concordia line. Because the two Lead isotopes would leak at the same rate (no chemical discrimination here), the ratios would both decrease— the amount of Lead would be lower than expected but the amount of Uranium would be unchanged. Because ^{238}U decays more slowly, the missing ^{206}Pb would be a larger fractional effect and the measured values would fall further below the concordia line. With a variety of Zircon samples that have lost different amounts of Lead, one can plot on the same diagram a "discordia" line. The point where the two lines intercept at their upper ends gives an estimate of the true sample age.

Just a few years after the discovery of the Chicxulub crater, T. E. Krogh, S. L. Kamo, and B. F. Bohor[8] measured the U-Pb age of Zircon crystals found at the K-Pg boundary layer in Colorado. These crystals showed large deformations attributable to the shock waves from the collision. They found an age of 65.5 ± 3.0 Mya, consistent with earlier determinations using Potassium-Argon (see below) and Rubidium-Strontium (see chapter 14) techniques. They also determined that the original formation of these Zircons crystals, excavated by the collision from perhaps 20 miles down in the Earth's crust, occurred 545 Mya. From plotting the discordia line, they conclude that some Lead was lost at the time of the ejection of these zircons. They attribute this loss to the heating of the crystals in the fireball that accompanied the impact, which blasted them to high altitudes before ultimately depositing them thousands of miles away.

Another venerable and ultimately more precise radioactive dating technique employs the decay of the element Potassium. Potassium (element symbol K) has two stable isotopes (^{39}K and ^{41}K) and one naturally occurring radioactive isotope ^{40}K with a half-life of 1.25 billion years. There are two decay modes, inverse beta decay and electron capture (see chapter 6):

$$^{40}K \rightarrow {}^{40}Ca + e^+ + v_e \text{ (89 percent)}$$

and

$$^{40}K + e^- \rightarrow {}^{40}Ar + v_e \text{ (11 percent)}$$

Argon is one of the noble gases that eschews chemical reactions of any kind and, at temperatures on Earth, is always in the gaseous state. When a rock is in a molten state—either in the form of hot lava from a volcano or melted by the heat of a collision—the gas formerly trapped in the rock is free to escape, but as the rock solidifies and locks all its atoms in place, each of the ^{40}K electron-capture events results in a lonely gaseous Argon atom locked in a cage of surrounding rock molecules. In principle, this allows us to apply the straightforward "accumulation clock" method described in chapter 6. Simply counting the number of ^{40}K and ^{40}Ar atoms and taking the ratio allows us to identify a unique point on the parent (daughter) decay (accumulation) curve (see figure 6.6) and read off the age directly. In practice, however, this involves some assumptions and a good deal of care in taking the measurements.

First, as noted in chapter 11, Argon is the third most abundant element in the Earth's atmosphere, and we must assume no air diffuses into the mineral as it cools from a liquid to a solid; otherwise, the initial ^{40}Ar abundance would not be zero, as required by the accumulation clock technique.[9] Furthermore, the rock crystal to be studied must be carefully chosen to have not suffered any fracturing so that some radiogenic Argon could have escaped. Thus, careful processing is required. The rock is crushed and individual crystals are selected by hand. They are heated gently to drive off any surface Argon from the air. Then they are placed in a vacuum furnace to which a precisely known amount of ^{38}Ar is added. The sample is heated to melting so that all the trapped gases (H_2O, CO_2, Ar, etc.) are released. The extraneous gases are frozen out with liquid nitrogen and the remaining Argon is sent to a mass spectrometer that separates the isotopes. Any ^{36}Ar that is present can only come from the air (where it makes up 0.334 percent of total Ar), so its abundance provides an absolute value for the air correction, allowing us to subtract the appropriate amount of each isotope. Thus, the amount of ^{38}Ar left is just the amount added for calibration purposes, and the ratio of this to the ^{40}Ar yields the number of radiogenic atoms in the sample.

Recently, a new version of this basic technique has been developed that makes the process somewhat simpler and reduces the uncertainties in the measurement. This takes advantage of the fact that there is another familial relationship between Potassium and Argon through the reaction

$$^{39}K + n \rightarrow {}^{39}Ar + p$$

Now, ^{39}Ar beta decays back to ^{39}K with a half-life of only 269 years; even for a young rock 10,000 years old, only one in 100 billion ^{39}Ar atoms will be left; for a 66-million-year-old sample, the amount remaining is precisely zero. In this method, one first bombards the crystal with neutrons to produce ^{39}Ar and then measures the relative amounts of ^{36}Ar, ^{38}Ar, ^{39}Ar, and ^{40}Ar, which allows for a simultaneous estimate of the air contamination and the amount of Potassium present, yielding the rock's age.

Tektites are millimeter- to centimeter-sized glassy globs produced in response to the extreme heat induced by a shock wave from an asteroid impact. They are found scattered around the Caribbean area in rock layers associated with the K-Pg boundary. P. R. Renne and collaborators used K/Ar dating on a collection of tektites from Haiti to establish the most precise date yet for the Chicxulub event at 66.038 ± 0.04 Mya, a remarkable accuracy of 0.06 percent (or only 32,000 years—equivalent to remembering when you woke up this morning to within 20 seconds).[10] They went on to compare this date with those derived from clay crystals called bentonites from Hell's Creek, Montana, that are coincident with the Iridium spike at the K-Pg boundary and found 66.043 ± 0.04 Mya, completely consistent with the tektite results from 3,000 miles away. The date of the dinosaurs' demise is one of the most precisely known in all of geology.

If this precision were not sufficient, we even have evidence of the month that the impact occurred. The Teapot Dome formation in Wyoming, famous because of the eponymous scandal in the Harding administration, contains all the geologic signs—enhanced Iridium, shocked minerals, tektites, and so on—of other K-Pg sites around the world. At the time of the event, the area was covered by a large, shallow inland sea. Jack Wolfe recovered fossils of lily pad leaves from the sediment layer coincident with the K-Pg boundary and showed they had distorted, fractured veins, an effect he managed to reproduce by sticking modern lily pad leaves in his freezer. He speculated that the "impact winter" (see below) induced by the stratospheric debris produced by the collision caused the plants to freeze. Given the reproductive stage at which the various species of water plants had reached, he concluded the impact must have occurred in early June.[11] For a minute-by-minute account of that fateful day, read on.

RAPID CLIMATE CHANGE

Three times in the last 100 years, major volcanic eruptions (Agung in 1963, El Chichon in 1982, and Pinatubo in 1991) threw sufficient material high enough into the stratosphere that it encircled the globe and caused global temperatures to fall from 0.10° to 0.25° C for a few years after each event. These explosive eruptions each ejected more than 10 cubic kilometers (approximately 25 billion tons) of matter. As discussed in chapter 11, even larger eruptions have occurred in historic times, such as the event in 1257 AD. That Indonesian eruption is estimated to have ejected 40 cubic kilometers of material and caused an agricultural collapse the following summer during which the sun was never seen. The Chicxulub impact is estimated to have produced 2,500 times as much: 100,000 cubic kilometers of ejecta.

And it hit in a most unfortunate spot for the dinosaurs, a site rich in anhydrite ($CaSO_4$) and its water-saturated form we call gypsum. This sent approximately 300 billion tons of sulfur into the stratosphere, where it combined with water to make sulfuric acid H_2SO_4 and condensed into tiny, highly reflective droplets that, combined with massive amounts of dust and soot, reduced the sunlight reaching Earth's surface by 80–90 percent for a decade or more. Photosynthesis essentially stopped, and the air temperature plummeted. The large heat capacity of the deep oceans, however, resists large, sudden temperature changes; the oceans stayed warmer than the air, leading to the conditions perfect for forming huge, violent storms that in turn kept the dust particles in the air even longer.

While the abrupt die-off of marine species at the K-Pg boundary is a clear signature of sudden dramatic change, the result is that fossils such as the foraminifera we used in chapter 11 to record ocean temperatures all died and are thus unavailable. J. Vellekoop and collaborators have recently circumvented this limitation and obtained a direct record of sea-surface temperatures in rocks from the Brazos River valley in Texas about 750 miles from the impact.[12] They extracted glycerol dibiphytanyl glycerol tetraether (GDGT) lipids that are produced by marine Thaumarcheota. These organisms are members of the Archaea kingdom, single-celled creatures that lack a nucleus and represent one of the three main branches of life on Earth. The complex organic fatty acids Thaumarcheota produce form the Archaea cell membranes. It turns out that their molecular structure is acutely sensitive to the temperature of their environments—they respond within hours to any temperature change. Extracting these compounds from the K-Pg boundary layer, the authors found a sudden drop from a mean sea-surface temperature of 30°C to 23°C (a 13°F temperature drop) within 1 cm of sediment. The temperature then returns to its earlier value about 20 cm above the boundary layer. This dramatic temperature transition is wholly consistent with an "impact winter" brought on by the asteroid impact.

After reaching its preimpact value, the temperature keeps rising to 32.5°C (over 90°F) 85 cm above the boundary in the sediment sample before drifting back toward the 30°C value. This is also consistent with one of the expected consequences of the collision. Vast quantities of water vapor were cast into the atmosphere when the impact fireball vaporized much of the Gulf of Mexico, as was more than 1 trillion tons of CO_2 from the calcite ($CaCO_3$) and other minerals and hydrocarbons that made up much of the rest of the geology at the impact site. This sudden input of greenhouse gases led to the observed temperature rise that probably persisted for at least a millennium. Since the dawn of industrialization in the late eighteenth century, the human input of CO_2 into the atmosphere is a comparable amount (approximately 1 trillion tons), and the observed temperature rise at the K-Pg boundary is similar to that predicted by current climate models for the end of this century

^{87}SR/^{86}SR IN SEAWATER

Another interesting signature of the asteroid collision comes from an analysis of the isotopes of Strontium (element number 38), which has four stable isotopes—^{84}Sr (0.56 percent), ^{86}Sr (9.86 percent), ^{87}Sr (7.00 percent), and ^{88}Sr (82.58 percent). The ^{87}Sr abundance is actually growing (very slowly) with time because it arises from the beta decay of Rubidium-87, which has a half-life of 48.8 billion years. (This means only 6.3 percent of ^{87}Rb has decayed since the Earth formed 4.5 billion years ago, so when I say slowly, I mean *really* slowly.) Rubidium-Strontium dating will become important in future chapters, so it is worth spending a bit of time to explore the relationship of these two isotopes. For the current application, it is important to note that Strontium lies directly under Calcium in the Periodic Table and thus is chemically similar, making it a ready substitute for Calcium in the little $CaCO_3$ shells of the foraminifera.

Of the Earth's crustal material, Strontium makes up about 370 parts per million (ppm), meaning ^{87}Sr is present at 25.9 ppm (7 percent of the total). Rubidium is rarer than Strontium, at 90 ppm, and the ^{87}Rb isotope is 27.83 percent of the total today; when the Earth formed, given the fraction that has decayed to date, it would have been 29.16 percent of the 91.68 ppm of Rubidium present at the time (the only other isotope, ^{85}Rb, is stable so it does not change with time). This means 1.68 ppm of today's ^{87}Sr comes from decayed ^{87}Rb. Even in a million years, the additional ^{87}Sr added by ^{87}Rb decay will amount to only .00035 ppm, a trivial amount. Thus, any sudden change in the ^{87}Sr/^{86}Sr ratio must have some external cause.

In seawater, the Sr and Rb concentrations are quite different from those in the rocks on land: Sr is present at 8.1 ppm and Rb at 0.12 ppm. But the ocean is a very big place; for example, this implies that there are 11 trillion tons of Sr in total and specifically 0.77 trillion tons of ^{87}Sr. Thus, to change the overall ^{87}Sr/^{86}Sr ratio requires a huge addition of one isotope over the other. The ^{87}Sr/^{86}Sr ratio in seawater today is 0.7091, while the ratio contributed to the sea by erosion of land-based rocks is closer to 0.716. Thus, any significant increase in the oceans' ^{87}Sr/^{86}Sr ratio implies a massive increase in erosion.

And that's precisely what several authors claim when examining the Sr ratios at the K-Pg boundary. From examination of the foraminifera shells in the rock strata, there appears to be a gradual increase from ^{87}Sr/^{86}Sr = 0.7077 10 million years before the boundary to 0.7078 at the boundary.[13] This has been explained as the gradual growth in land area caused by plate tectonics over this period (more land means more opportunity for rock erosion). At the boundary itself are several measurements, and they do not agree with each other very well, although all appear to show a small sudden increase—estimates range from a jump of .0002[14] to only a quarter of that amount. But, again, using the numbers above, even the lower change of 0.00005 in the ratio requires an influx of about 5 billion tons of extra ^{87}Sr. Given the modest enhancement of ^{87}Sr in land-based rocks from which this excess presumably came, we require a total of 65 billion tons of Sr from eroded rocks or 175 trillion tons of rocks in total.[15]

That sounds like a huge amount of sudden erosion, but the Earth is a big place. The continental crust has a total mass of 2×10^{22} kg, so this represents only 0.001 percent of the total (equivalent to half an inch off a mile-thick layer of rock). But what would cause such a sudden change in erosion rates? Two consequences of the asteroid strike are obvious culprits.

First, the shock wave sent round the world by the collision would have converted a lot of the N_2 in the air to nitric acid (HNO_3)—about 60 billion tons of it. As noted above, the anhydrite at the site of the collision thrust an enormous quantity of what ultimately became sulfuric acid (H_2SO_4) high into the stratosphere. We know from recent volcanic eruptions that these acids produce acid rain for several years as they gradually settle out of the atmosphere. One estimate is that each square meter of Earth received several liters of highly acidic rain, sufficient to increase the erosion rate and explain the spike in the ^{87}Sr/^{86}Sr ratio.

OTHER IMPACT CLUES

A host of other clues support the scenario of a catastrophic impact. All around the Gulf of Mexico are deep sandstone layers near the K-Pg boundary layer representing compressed layers of sand deposited inland. Careful examination of these deposits shows that, apart from the topmost layer, there are none of the small worm tracks that are ubiquitous in sandstone deposits of all ages around the world. This supports the notion of very rapid deposition, as would be expected from enormous tsunamis (up to a mile high) generated by the impact. In this case, the Iridium spike is in a mud layer just on top of the sandstone. This is explained as the result of the slower settling of the fine-grained mud and its associated pulverized asteroid material, as distinct from the heavier sand grains which settled out first.[16]

At sites around the world, from Europe to New Zealand, soot from burned organic matter is mixed with the clay that marks the K-Pg boundary. Wolbach and colleagues estimate the total amount of combusted carbon at 70 billion tons, comparable to 10 percent of all the plant material present on Earth today.[17] This evidence for a global forest fire is supported by the fact that there is a sudden decrease in the fraction of ^{13}C in marine shells across the boundary. Recall that plants discriminate against ^{13}C in photosynthesis; thus a sudden release of CO_2 from burning plants will lead to a decrease in the atmospheric and oceanic ^{13}C fraction (much as we see today from the burning of fossil fuels). There is also a sudden increase in fern spores above the K-Pg boundary, indicative of land cleared by fire that is regenerating plant life (which often begins with fernlike species).

Other authors have argued against global destruction, suggesting the soot arose from massive fires associated with burning of fossil carbon (coal and oil) at and near the impact site (recall it is at a site where Pemex was drilling for oil). The putative lack of charcoal in the boundary layers is another argument used against the global wildfire hypothesis. However, reanalysis of the charcoal particles present in the layer and detailed modeling of both the initial fireball (which could have directly ignited all plant material out to a radius of 1,500 miles) and the fallback of debris from the upper atmosphere, which could start fires anywhere in the world, suggests that, while the whole world did not go up in smoke at once, fires were common everywhere.[18] Indeed, the months of darkness following the collision likely killed a lot of the plant life, making it more vulnerable to forest fires started by lightening.

Recently, a rich cache of fossils was discovered in North Dakota that shows direct evidence of a cataclysmic event that erased life in a matter of minutes. The authors describe their find as follows: "A tangled mass of freshwater fish, terrestrial vertebrates, trees, branches, logs, marine ammonites and other marine creatures was all packed into this layer . . . We have one fish that hit a tree and was broken in half." Some of the fish were found to have tiny glass spherules ejected from the impact caught in their gills.[19] The Iridium-rich dust layer is found just on top of these fossils. The authors posit that the mashup of fossils was caused by a tsunami-like wave of water called a seiche. From both the fossil deposit and scaling from seiches known to be caused by earthquakes today, they estimate the wave was more than 10 meters (30 feet) high and was induced by the magnitude 10 to 11.5 earthquake representing the shock wave from the impact 1,850 miles away that arrived between six and thirteen minutes after the event. The larger impact-produced glass spherules and other debris began arriving roughly fifteen minutes after the collision and are thus found on top of the fossil layer. The Iridium-enriched fine dust settled out yet later, as indicated by its presence on top of the ejecta material. This find provides a vivid picture of the chaos that followed this multimillion megaton event.

From a casual observation of an abrupt change in the marine fossil record in 1860 (Phillips), to a geochemical exploration of this boundary in 1980 (Alvarez et al.), to a direct record of chaotic destruction in 2019 (DePalma et al.[20]), the dramatic story of the last mass extinction has emerged. Our atomic historians have allowed us to determine the date of the catastrophe, the location of the devastating impact, and the sequence of events from minutes after the asteroid hit to a millennium later as life was reestablishing itself on land and in the sea. That life, devoid of the dinosaurs that had ruled the Earth for 200 million years, left an ecological niche in which furry little mammals could flourish and, 66 million years later, give rise to a species that could reconstruct this dramatic history in such exquisite detail.

CHAPTER 13

Evolution

From Meteorites to Cyanobacteria

A s we argued in chapters 3 and 4 the ninety-four elements we find on Earth are identical to those found throughout the universe. All the stars, planets, and clouds of gas and dust are made of exactly the same stuff, albeit in different proportions. If I were to take a fair sample of matter from throughout our galaxy, I would find that, for every million atoms, 923,346 of them were Hydrogen, 74,980 of them were Helium, 812 were Oxygen, 470 were Carbon, and the other ninety elements were all less than 100 parts per million (ppm), that is, less than 0.01 percent. The reason for this radically skewed distribution is revealed in chapter 16.

In our solar system, things are just as heavily weighted toward the lightest two elements: H + He make up 99.86 percent, with Oxygen and Carbon coming in at 477 and 326 ppm, respectively; Nitrogen (102) and Neon (100) just sneak over the 100-mark. This is unsurprising because our solar system formed from the raw material of the galaxy.

On the surface of the Earth, however, things are very different. Virtually all the original Hydrogen and Helium atoms have escaped—the lightest particles move fastest (see chapter 3 and below) and, in this case, fast enough to leave the Earth entirely unless the H is bound with Oxygen in water or with some other atoms in a gas or a mineral. That leaves Oxygen as the most common element in the Earth's crust (467,100 ppm) followed by Silicon (276,900 ppm), Aluminum (80,700 ppm), Iron (50,500 ppm), Calcium (36,500 ppm), Sodium (27,500 ppm), Potassium (25,800 ppm), Magnesium (20,800 ppm), Titanium (6,200 ppm), Hydrogen (1,400), and Phosphorus (1,300) with everything else coming in at less than 1,000 (0.1 percent). Note that Carbon, the fourth most abundant element in the universe, is present at just 900 ppm (0.09 percent).[1]

Because 71 percent of the Earth is covered by the ocean, we should complete this inventory by specifying the abundances of the elements in seawater: Hydrogen (66,444 ppm), Oxygen (32,945 ppm—a ratio of roughly 1:2 with Hydrogen because almost all of it is H_2O), Chlorine (3,400 ppm), Sodium (2,100 ppm—sodium chloride (NaCl) is one of the salts in salt water), Magnesium (330 ppm—MgCl is another salt), Sulfur (170 ppm), and everything else less than 100, including Carbon at only 19 ppm. Carbon, the "building block of life," is the fourth most common element in the universe, but it is rather sadly depleted on Earth's surface.

THE CHEMICAL PRECURSORS

In addition to the fact that the physics of all atoms—their masses, isotopes, electron energy levels, and so on—is identical throughout the cosmos, their chemistry (attraction and aversion to other atoms, bond strengths, etc.) are also identical. In fact, we have detected over 200 molecules in interstellar space and in the outer atmospheres of stars cooler than the Sun, with the largest molecules containing up to seventy atoms. All are molecules also found Earth.[2] Most of these molecules are made from the universe's three most abundant elements capable of forming bonds: Hydrogen, Oxygen, and Carbon. Indeed, 75 percent of the three- and four-atom molecules contain Carbon, as do all but one of the eighty-eight larger molecules (the exception is SiH_4). Many of the molecules in space are common on Earth, including water (H_2O), formaldehyde (H_2CO), ammonia (NH_3), ethanol (CH_3CH_2OH), acetone (C_2H_6CO), and urea (N_2H_4CO). Earth is chemically at one with the universe.[3]

Despite all these atomic and molecular similarities, there is one thing about Earth that is, so far at least, unique. Earth hosts life.

If the raw ingredients are the same everywhere, and the recipes for their combinations are likewise identical, what's the special sauce? Why is our pale blue dot the only place life has emerged?

We don't know, of course, that we are unique. There are several places close to home in the solar system where the emergence of life is possible: the subsurface of Mars, the warm salty oceans under the icecaps of Jupiter's moon Europa and Saturn's moon Enceladus, or the lakes of liquid methane on Titan. None of these has been searched carefully to date.

It has been only twenty-five years since we discovered that our solar system is not unique—a majority of the hundreds of billions of stars in our galaxy have planetary systems. We've confirmed over 5,000 planets as this book goes to press, and more are being discovered every day. Some of the new worlds are very similar in size and density to the Earth, and a subset of those lie at distances from their parent stars where Earth-like temperatures are expected to prevail. We have even detected water and carbon dioxide in the atmospheres of some of these planets, although studying them in detail is very challenging: observing an Earth-like planet orbiting a Sun-like star in the solar neighborhood only 100 light years away is like trying to study a 0.5-watt Christmas tree bulb next to twenty times the illumination of New York's Times Square—from Washington, DC. Finding microbes on such a planet is beyond our current capabilities, although should they be numerous enough to alter the composition of their planet's atmosphere, as they have done on Earth, we can expect to discover this in the coming decade.

THE OLDEST LIFE

Currently, however, we are stuck with this single example of a planet containing life. How quickly did it form? And how long did it take to evolve creatures who could ask that question?

As our atomic clocks will reveal in chapter 14, the Earth is 4.567 billion years old. Because of the constant motion of the continental plates colliding and separating and remaking the landscape, there are no rocks this old on the surface today (at least that we have found—we date the Earth's formation from meteorite ages, as described in chapter 15). Indeed, the oldest crystals found on Earth are zircons (discussed in chapter 12) from Western Australia. One deep purple zircon only 0.2 mm across was found to have a U-Pb age of 4.404 ± 0.008 billion years. The fact that this crystal's $^{18}O/^{16}O$ ratio was high has been used to infer the presence of abundant liquid water at the time of its formation. Thus, less than 170 million years after the planet's formation (only 4 percent of its lifetime to date) we had a solidifying crust and oceans.[4]

These tiny zircon crystals have survived the destruction and recycling of the rocks in which they were formed. The oldest complete rocks in the continental crust were found about a decade ago in the northern part of Quebec in Canada. The dating technique used two of the rare earth elements, Samarium (Sm) and Neodymium (Nd). Element number 62, Samarium has five stable isotopes plus two very long-lived radioactive ones: ^{147}Sm ($t_{\frac{1}{2}}$ = 105 billion years) and ^{148}Sm that lives 75,000 times longer (so long it is largely indistinguishable from stable). ^{147}Sm undergoes an alpha decay to ^{143}Nd. Since the origin of the solar system, only 2.9 percent of this isotope has decayed, but that's enough to increase the amount of ^{143}Nd in meteors found today compared to younger Earth-bound rocks in which the Sm has been trapped for less time and has thus produced less Nd. The process for determining the rock age uses "isochrones," a slightly complicated procedure we will explain in chapter 14, as we discover the date of the solar system's birth. But the result of applying this radioactive chronometer to the igneous rock outcrops in northern Quebec yields an age of 4.3 to 4.4 billion years, only slightly younger than the Australian zircons and further evidence that the Earth was then—only approximately 0.2 billion years after its formation—a recognizable place.[5]

How much longer did it take for life to emerge? As one might expect, the quest to find the oldest evidence of life is both highly competitive and contentious. As this book goes to press, the oldest claim derives from analyzing the Carbon isotope ratio in graphite found embedded in old rocks from Newfoundland in Canada. Uranium-Lead dating of zircons in this outcrop that has been metamorphosed from sedimentary rock indicates an age of 3.95 billion years. The researchers extracted Carbon from carbonate rock and from graphite found embedded in the rocks. They find the carbonate has a ^{13}C/^{12}C ratio more than 2.5 percent greater than the ratio for the graphite (−0.26 percent versus −2.82 percent compared to the standard value). As we showed in chapter 10, the chemical incorporation of Carbon from the environment into living things systematically discriminates against the slower moving, heavier ^{13}C. After considering all the geologic processes these rocks have gone through over their nearly 4 billion years, the authors conclude that the graphite must have been produced by single-celled organisms that extracted nutrients directly from their environment (so-called autotrophs).[6] Far from everyone agrees with this interpretation.

Less controversial is the evidence from stromatolites. These layered rocks arise from multiple layers of photosynthesizing bacteria that create large mats containing trillions of organisms. Cyanobacteria are thought to be responsible for creating much of the Oxygen in the Earth's atmosphere (see below), and they are seen producing such structures today where they live in salty environments (e.g., Great Salt Lake and similar highly saline lakes as well as marine lagoons).

Fossilized stromatolites in Australia have variously been dated to between 3.4 and 3.7 billion years ago, and in some instances also show the ^{13}C deficits expected from photosynthesizing organisms.

It appears that life not only arose but was abundant enough to leave fossil evidence sometime between 3.5 and 4 billion years ago. This is quite rapid, considering that, for the first 100 to 200 million years, Earth was still being bombarded by rocks from space and cooling from its initial molten state. As we will see in chapter 16, the Sun will remain relatively unchanged for a total of 10 billion years, meaning that life arose within only approximately 6 to 8 percent of the Sun's lifetime. In one sense, this is unsurprising because we know that all the raw materials (the organic molecules cited above) are present in precisely the clouds of gas in interstellar space where new solar systems are born. The missing step, of course, is how we get from those raw materials to self-replicating organisms that store information in their DNA molecules and pass it on to new generations—with the occasional error that leads from a cyanobacterium to you.

LEFT-HANDED LIFE

By now, we are comfortable with the fact that each of the basic building blocks of matter—the elements—comes in a variety of flavors called isotopes. Indeed, we have seen how remarkably useful these varieties are in reconstructing history. The basic building blocks can be combined in millions of different combinations—the molecules—each of which has specific properties that depend on the exact combination of atoms involved. It turns out that some molecules, even when composed of identical combinations of atoms, can also come in different flavors, which we label right-handed and left-handed. This concept is called chirality.

Your two hands are, I suspect, remarkably similar to each other. But that doesn't mean you can put your right hand into your left glove; no matter how you rotate or contort it, your right hand won't fit. Your two hands are mirror images of each other. Likewise, two molecules with exactly the same chemical formula—the same number of each kind of atom with all the same chemical bonds—can be mirror images. One of the simplest examples is butanol ($C_4H_{10}O$), a colorless flammable alcohol used in industrial processes to make plastics and lubricants (see figure 13.1).

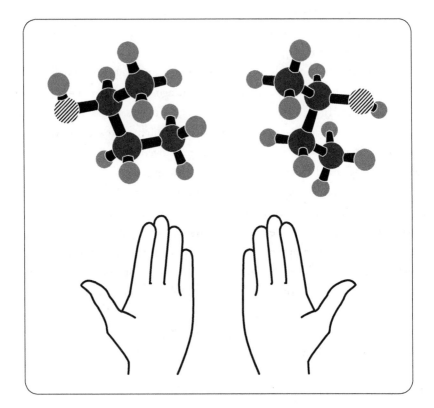

13.1 Right- and left-handed isomers of the butanol molecule. While the two molecules have exactly the same chemical formula, it is not possible to rotate one to make it identical to the other; like your two hands, they are mirror images of each other (note the location of the striped atom of Oxygen).

In biological processes, the exact shape of a molecule is everything. It determines how the molecule interacts with its neighbors and how it behaves as it moves through the world. The thousands of different proteins in each of your cells all have very specific shapes that determine their functions. As they work, they can fold into complex conformations that are determined by the exact arrangement of their chemical bonds. Chiral molecules, like our hands, truly have different shapes, and thus the left- and right-handed ones have different properties. For example, the left-handed version of carvone ($C_{10}H_{14}O$) smells like spearmint, and the right-hand version smells like caraway. Their different shapes mean that they lock onto and stimulate different receptors in your nose and thus send different signals to your brain. For aspartame, the artificial sweetener that is 200 times as sweet as sugar, only the left-handed version of the molecule is sweet; the right-handed version is tasteless—again, different shapes imply locking onto different receptors that send different signals to the brain.

Many organic compounds are chiral. In using the drug penicillamine ($C_5H_{11}NO_2S$) to treat acute rheumatoid arthritis, among other conditions, one must be careful to select and administer only the right-hand molecules because the left-hand version is toxic—it blocks the action of the essential vitamin B_6. Perhaps the most infamous and tragic case of drug chirality is thalidomide. Marketed in Germany in the late 1950s for the relief of morning sickness in pregnant women, it caused between 5,000 and 10,000 severe birth defects and hundreds of babies' deaths.[7] Only the left-handed molecule causes these problems, but because this was unknown at the time, the drug was sold in what is called a racemic mixture—equal amounts of right- and left-handed molecules (an equal mix is what emerges from a typical chemical synthesis process for any molecule).[8]

It turns out that we—and I am using that term as inclusively as possible here meaning *Homo sapiens*, bacteria, and everything in between—are made of chiral molecules. And "we" are all left-handed.

Proteins are the workhorse molecules of life. They provide cells and organs with structure; replicate the DNA every time a cell divides; act as catalysts in chemical reactions; extract, store, and utilize energy; trigger and respond to stimuli; transport molecules around inside cells and in intercellular spaces. The roughly 20,000 genes in your DNA each provide a unique code to build one of these proteins. The code specifies the ordering of the protein's building blocks, the biological molecules called amino acids. All life on Earth uses just twenty different amino acids,[9] and the chiral ones are all left-handed.

Each of the amino acids are composed of Hydrogen, Nitrogen, Oxygen, and Carbon, with two of the twenty adding a single Sulfur atom. They range in size from ten atoms (Glycine: $C_2H_5NO_2$) to twenty-seven atoms (Tryptophan: $C_{11}H_{12}N_2O_2$). The smallest protein known, dubbed TRP-Cage and consisting of a chain of only twenty amino acids (less than 500 atoms), was discovered in the saliva of Gila monsters.[10] Don't ask me why Gila monster spit was a subject of study, but because this tiny protein has become invaluable in studying how proteins fold and assume their functional shapes, I'm glad someone decided it was—the power of curiosity-driven science must never be underestimated. The largest protein known is Titin, critical to muscle contraction.

It is comprised of 34,500 amino acids (over half a million atoms), with 244 individual folding domains.

All these folds—in TRP Cage and Titin alike—happen in precisely the way they do because all the chiral amino acids are left-handed. Because the folding pattern is critical to the protein's function, this single-handedness of the building blocks is essential. If the little protein factories called ribosomes in the cell had an equal number of right- and left-handed amino acids to choose from, they would randomly grab roughly half of each, and every protein they made would be nonfunctional—it simply couldn't fold into the shape required for its task. While there is nothing special about left versus right, what is essential is that, in any given organism, only one must prevail.

WHENCE CHIRALITY?

The origin of this ubiquity of left-handed amino acids is the subject of much debate. The simplest hypothesis is random chance. When the first self-reproducing molecule formed in the primordial soup, it happened to select left-handed amino acids from an otherwise 50:50 mixture of left- and right-handed molecules—like flipping a coin—and the rest is history. One corollary attendant on this hypothesis, however, is that the formation of life is rare—that it happened only once on this large planet over its entire 4.567-billion-year history. Otherwise, if it happened several times, we should expect some species descended from a left-handed line and others that use right-handed molecules.

There is a fascinating clue found in meteoritic material that suggests an alternative hypothesis for our left-handedness. In September 1969, a bright fireball was seen in the sky near the small town of Murchison in Victoria State, Australia, about 100 miles north of Melbourne. A total of more than 100 kg in pieces as large as 7 kg (15 pounds) were collected from this meteorite. The meteor was of the relatively rare class (less than 5 percent) known as Carbonaceous Chondrites, which are thought to be among the most primitive bodies in the solar system. They contain many Carbon molecules and also usually water.

The Murchison meteorite is among the best studied in the world. When abundant amino acids were first identified in the object (at the level of 100 ppm), controversy ensued over whether they were simply contamination from Earth after it landed (e.g., picked up from the ground on which the pieces lay or from subsequent handling). The measurement of the Carbon isotope ratio $^{13}C/^{12}C$ resolved this concern: the ratio in the nucleotide Uracil ($C_4H_4N_2O_2$) was more than 5 percent above that in the atmosphere, compared to the 2 percent below atmospheric composition typical of terrestrial organisms because of life's discrimination against the heavy isotope (see chapter 10). In fact, a total of more than seventy amino acids have now been identified in the meteorite, including eight of the twenty found in living things, along with over 14,000 other organic compounds.[11]

Most interesting in the current context, however, was the discovery from the meteoric material that the amino acid alanine ($C_3H_7NO_2$) had an excess of left-handed molecules. While initially reigniting the contamination hypothesis, it was subsequently shown that isovaline ($C_5H_{11}NO_2$), which is not one of the twenty amino acids of life, also showed an excess of left-handed molecules, strongly suggesting some nonbiological process led to this preference. A recent meteorite fall in Costa Rica also contained isovaline, in this case with a 15 percent excess of left-handed molecules.[12]

To usher in the new millennium in January 2000 (albeit eighteen days late), a fireball lit up the morning sky over the border of British Columbia and the Yukon Territory in northwest Canada. The estimated size of the incoming meteor was 66 tons, although all but about 2 percent is expected to have been burned off in its plunge through the atmosphere before it hit the ice of Lake Tagish. About 10 kg (22 pounds) of fragments were ultimately collected (only 1 percent of the total that landed). The amino acids threonine, serine, aspartic acid, and glutamic acid were found to have astonishing left-handed excesses of 99 percent, 80 percent, 45 percent, and 50 percent, respectively.[13] The alanine, however, was only enriched in left-handed molecules by 8 percent (although, as always, it was a left-handed excess). These amino acids had strong ^{13}C enhancements, ruling out contamination. Laboratory studies of aspartic acid show that it can make crystals that are either left-only, right-only, or an equal mixture of left- and right-handed molecules, and that different temperature and concentrations affect the outcomes differently.[14]

The left-handed acids always seem to be more abundant, although most of the excesses are modest, with the exception of the Tagish Lake samples. Is this small bias enough to ensure that all life is left-handed? In 2008, my late Columbia colleague Ron Breslow, former president of the American Chemical Society, delivered a talk to the society's annual meeting in which he reported on how a modest imbalance of left- versus right-handed molecules can be amplified by the conditions expected on the primitive Earth. Working with Columbia graduate student Mindy Levine (now a professor at the University of Rhode Island), he showed that adding amino acid precursors (pieces of the complete molecules) to a solution of left-handed molecules made more left-handed molecules. Not only that, by beginning with a solution containing just a 5 percent excess of left-handed amino acids, equal numbers of left- and right-handed molecules would crystalize together and precipitate out of solution, leaving behind a left-handed dominated population in the water.[15] The small excesses that the meteors delivered to Earth in its earliest days can thus come to dominate the precursors of life, cascading into the left-handed world we inhabit today, 4.5 billion years later.[16]

This begs the question: Why did the early solar system in which the meteors formed have a preference for left-handed molecules? There are several proposed explanations for this, but I have my favorite. As we will see in chapter 15, there is evidence that a star exploded in our vicinity not long before the solar system began to form. One product of such a stellar explosion is the collapsed core of the star, which turns out to have the density of an atomic nucleus but is 10 km across—a neutron star (see chapter 16). These astonishing objects rotate at incredible speeds (up to fifty times a second) and have a magnetic field 10 to 100 trillion times stronger than that of the Earth's. The result is that beacons of radiation pour out into space along their north and south magnetic poles. One of the properties of this radiation is that it is "circularly polarized"; that is, the orientation of the oscillating electric and magnetic fields we discussed in chapter 4 trace a circular pattern as they move through space.

The light from one of the neutron star's poles exhibits a clockwise rotation, while the other pole's light rotates counterclockwise. Circularly polarized light can excite oscillations of molecular bonds and, if the excitation gets too great, break them. Right-handed molecules respond more strongly to clockwise rotations. Thus, if a nearby neutron star's clockwise pole was pointed toward the cloud of gas and dust destined to become the solar system, it could have systematically destroyed more right-handed molecules than left-handed ones, thus explaining the bias that has led to globally left-handed life. While this dramatic picture is attractive, it is clear that it cannot account for the largest excesses seen in meteorites and that processing in the meteor's parent body and/or after its arrival on Earth is also required.[17]

In the foregoing, I have repeatedly referred to "life" as left-handed, but this only applies to the chiral examples among the twenty amino acids that are the essential building blocks of proteins. The code to build those amino acids is stored in the giant molecule deoxyribose nucleic acid (DNA). The spiraling backbone of the DNA molecule that holds in place the coding bits is a sugar molecule ($C_5H_{10}O_4$), and sugars, it turns out, are all right-handed molecules (i.e., the DNA backbone spirals to the right). Does this put a dent in my elaborate argument above on the origin and amplification of left-handed amino acids?

No, it doesn't. In 2004, Sandra Pizzarello from Arizona State University combined the left-handed excess of isovaline seen in the Murchison meteorite with glycoaldehyde and formaldehyde, two molecules thought to be prevalent on the newborn Earth. The resulting chemical reaction produced threose ($C_4H_8O_4$), a somewhat simpler sugar than the deoxyribose of DNA but one that scientists have speculated could have been a precursor to the self-replicating DNA molecules that now form the basis of life. The threose produced from Pizzarello's mixture showed a 5 percent excess of *right*-handed molecules, perhaps the precursors of the right-handed world of sugars that predominate today.[18] In addition, there is a recent report of the detection of ribose (the RNA backbone sugar) in the Murchison meteorite, and it too shows a large excess of right-handed molecules.[19]

Despite all these tantalizing clues concerning the origin of life on Earth, we remain far away from a complete picture of the process. As noted above, we know it happened quite rapidly, but we don't know how rapidly. We know it formed at least once, but we don't know if it happened more than once at more than one location (a critical consideration when speculating about the prevalence of life elsewhere in the universe). We also don't know how the crucial first step occurred—the transition from chemical reactions that occur throughout the cosmos to biochemical reactions, unique so far to Earth.

EVOLUTION SINCE THE BEGINNING

Life on Earth today can be divided into three distinct kingdoms: the Bacteria and the Archaea, both classes of single-celled organisms that lack a cell nucleus, and the Eukaryotes, including everything else, from single-celled diatoms to humans. Much effort has been expended to describe and date the ultimate ancestor, called the last universal common ancestor (LUCA),[20] primarily by attempting to identify the genes most common throughout the kingdoms and then, inferring from the protein products they encode, what that single-celled organism was like. Unfortunately, there is to date no ticking atomic clock or a unique chemical signature to help in this search.

One thing is clear, however. Soon after life was established, the magic of Darwinian natural selection meant that it rapidly began exploiting the wide variety of ecological niches available on the planet. And it soon made the Earth a fundamentally biological place in which life shaped the environment while the environment was shaping life. We'll explore just a few of these symbiotic interactions in the next chapter.

CHAPTER 14

What's Up in the Air?

Earth's Evolving Atmosphere

It is not random happenstance that life developed on Earth instead of on another planet in the solar system (assuming it hasn't). The elaborate and rather delicate molecules of life need a nurturing environment in which to form and evolve. The most important consideration is temperature. If it is too hot, the atoms and molecules are moving fast, and collisions between speedy molecules are likely to break them apart. If it is too cold, everything condenses into the solid phase, in which atoms are locked in place, so it is difficult for them to get around and find companions to hook up with to make molecules.

The "Energy in = energy out" equation introduced in chapter 11 allows one to estimate the temperature of any planet, given the output of its parent star and its distance from that star. The Earth is the one planet in the solar system in which that temperature estimate is in the Goldilocks range—neither too hot nor too cold. As noted in chapter 11, however, the calculation for a bare Earth gives an average temperature of 23°F, a little on the chilly side over most of the planet. It's our atmosphere blanketing the Earth that makes the climate ideal.

The Earth's atmosphere is also in the Goldilocks zone when it comes to the amount of air we have. Venus's atmosphere has 100 times the density, and the runaway greenhouse effect (see below) means its surface temperature is nearly 900°F. Mars has only 1 percent of the Earth's atmosphere and, coupled with its greater distance from the Sun, its surface temperature is −80°F. Our air, extending only 1 percent of Earth's radius above the surface and containing only 1 millionth of the planet's mass, is just right for creating the optimal conditions for complex molecules to form and evolve. It wasn't always so and, indeed, the existence of life itself has greatly modified our atmosphere's composition and continues to do so to this day.

THE PRIMORDIAL ATMOSPHERE

As noted in the last chapter, the solar system is predominantly Hydrogen and Helium. As a consequence, every planet began with an atmosphere dominated by these two elements along with the simple molecules Hydrogen can make with the next most abundant elements of Oxygen, Carbon, and Nitrogen: water (H_2O), methane (CH_4), and ammonia (NH_3). Note that the abundant Hydrogen joins each atom with enough friends to fill the heavier atom's missing outermost electron shell—two for Oxygen, three for Nitrogen, and four for Carbon, as seen in chapter 3. Today, the air on Earth is composed of 77.9 percent N_2, 20.9 percent O_2, and 0.9 percent Argon. The original five atoms and molecules together make up less than 0.25 percent of Earth's modern atmosphere, and most of that results from evaporation of H_2O from the oceans, which cover 71 percent of the planet. Helium is present at only 5 parts per million (ppm); methane is 1.9 ppm; H_2 is 0.5 ppm; and ammonia, detected in the air for the first time only in 2016, is 0.000033 ppm. What produced such a drastic change?

As described in chapter 3, all atoms are in motion, and temperature is just a measure of the kinetic energy of the atoms and molecules that make up a substance. The atmospheric particles on any planet move with a velocity characteristic of the temperature of that planet—on Earth, with a mean surface temperature of 16°C (58°F), the average speed is roughly 450 m/s (approximately 1,000 mph) for the N_2 molecules. The force that holds these fast-moving particles to Earth is gravity. The strength of gravity at the surface is determined by the mass and diameter of the planet. As demonstrated by the fact that we have sent rockets to the Moon, to the other planets, and even out of the solar system[1] it is possible to achieve a speed great enough to escape the bonds of Earth's gravity permanently; from Earth's surface, the escape speed is 11.2 km/s (about 25,000 mph). Any atom or molecule that achieves this speed will leave the planet, never to return.

The average speed of even the lightest molecule, Hydrogen, is only 1.8 km/s (about 3,900 mph), so it might seem as though no atoms or molecules could escape (at least without NASA's assistance). But recall from chapter 3 that, while many molecules cluster around the average speed, there are some moving much faster (that's how your dishes dry in the dish rack overnight). Thus, some of the original light H_2 began leaking away right from the start. In addition, there are ways other than collisions with their neighbors for particles to achieve high speeds. If atoms or molecules gain energy by absorbing high-energy ultraviolet (UV) solar photons or from collisions with high-speed particles spewed out by the Sun, a planetary wind can develop that sweeps lots of atmospheric matter into space, with the lightest particles carrying some of the heavier ones along with them, just as the wind can sweep up heavier particles of dust.

In its youth, our Sun was much more active than it is today. It was rotating much faster, and that rotation speed generated a stronger magnetic field, which in turn supported giant flares and a robust solar wind, wreaking havoc on the primordial atmosphere. In addition, constant bombardment of the Earth by massive asteroids within the first few hundred million years could vaporize entire oceans, driving clouds of steam into the upper atmosphere, where UV light could unbind H_2O molecules, allowing the Hydrogen to escape (see below). By 4 billion years ago, most of the primordial atmosphere was gone. Evidence for such escape is found in the isotope ratios of the noble gases; under this scenario, the lighter isotopes (and atoms) should escape more easily. Indeed, we find the Neon isotope ratio $^{20}Ne/^{22}Ne$ is on Earth 0.102, 16 percent below that of the original solar nebula from which the planets and their atmospheres formed. The ratio of Neon (atomic mass 20) to Argon (atomic mass 40) on Mars, Earth, and Venus is, in each case, only 1 percent of the solar value, indicating preferential escape of the lighter atoms.[2]

While the primordial atmosphere was being stripped from the planet, erupting volcanoes and continuing impacts were contributing molecules of Nitrogen (N_2), carbon monoxide (CO), carbon dioxide (CO_2), water (H_2O), and other molecules in smaller amounts. In the liquid water oceans, the CO_2 could react with Magnesium and Calcium to make carbonate rock, thus taking much of the CO_2 out of the air and beginning the Carbon cycle that remains active to this day (CO_2 in air → carbonates in ocean → limestone on ocean floor → limestone on land due to plate tectonics → weathering of rocks, volcanoes → CO_2 in atmosphere and oceans).

As we established in chapter 13, life flourished on Earth before the planet's first billion-year birthday, and the cyanobacteria began their work to transform the atmosphere again by pumping out Oxygen. For more than a billion years, this highly reactive element combined with elements on the Earth's surface and in the oceans (particularly with Iron as Fe_3O_4 and Fe_2O_3), leading to the banded iron formations, the source of much of today's Iron ore.[3] Eventually, the oxidation of Iron and other elements began to decline, and the O_2 produced by the photosynthesizing bacteria started building up in the atmosphere in what is called the "Great Oxidation Event" between 2.5 and 2.4 billion years ago, when atmospheric O_2 went from essentially zero to approximately 4 percent. This was fatal to many single-celled organisms that could only survive in an anoxic (i.e., Oxygen-free) environment, although some, such as methane-producing Archaea and the anaerobic botulism bacterium (*clostridium botulinum*), survive to this day.

The Oxygen level in the atmosphere remained at this level for more than another billion years, until about 650 million years ago (Mya) when it started to rise after having saturated all the metals with which it could combine both on land and in the sea. It rose rapidly to 12 percent in the next 100 Myr, until the Cambrian explosion—the sudden blossoming of multicellular plant and animal life in the ocean and the first plants on land. With land-based plants now contributing O_2 (and few land-based animals to breath it in), the level continued to rise, reaching a peak of 35 percent of the atmosphere about 280 Mya, when massive amounts of plant life were being buried each millennium to become the coal beds of today. Insects, which absorb Oxygen directly through their shells and so are limited in size by the Oxygen concentration, reached their apex, with dragonfly-like species having wingspans of over two feet.

By this time, amphibians and reptiles populated the Earth. Massive volcanic eruptions around 241 Mya caused a drastic decline in O_2, to about 15 percent (and led to the greatest mass extinction of the last half billion years). The Oxygen content then slowly recovered, getting back to a peak near 30 percent when the largest dinosaurs ruled the Earth and then declining after the K-Pg asteroid hit, reaching the 21 percent we experience today by about 25 Mya. Over the last several billion years, then, there has been a feedback loop in which the atmospheric composition influences the conformation of life and the presence of evolving life leads to an evolving atmospheric composition.

ATMOSPHERIC CHANGES: THE THERMAL MAXIMUM AFTER THE CRASH

As we saw in chapters 12 and 13, changes to the trace constituents of the atmosphere can also produce a large effect in the conditions at the surface of the Earth. Many studies have used signatures in the isotope record to document major changes in Earth's climate in response to geological and biological activity.

Here, I will highlight two examples, the period of pronounced warming that occurred about 10 million years after the meteor strike that wiped out the dinosaurs, and the end of the last ice age 12,000 years ago.

The Paleocene-Eocene Thermal Maximum (PETM) is one of the most dramatic sudden changes in Earth's climate since the Chicxulub impact. The event's signatures are a sudden rise in temperature coincident with an abrupt change in the $^{13}C/^{12}C$ ratio that indicates a rapid release of large amounts of Carbon to the environment. The date of this event has been narrowed down by the U-Pb dating of zircons from Spitsbergen Island in the Arctic Ocean as being between 55.73 and 55.96 Mya; a coincident date of 56.09 ± 0.13 Mya has been derived using the same method on zircons from Venezuela.[4]

At that time, surface-dwelling foraminifera show a sudden drop in their $^{13}C/^{12}C$ ratio, a simultaneous increase in the $^{18}O/^{16}O$ ratio, and a decline of 0.2 percent in the ratio of $^{11}B/^{10}B$. The first of these you will infer represents the addition of Carbon from a source with a different isotope ratio than normal, and the second we've used as a proxy for temperature. Element number 5, Boron, is a new entry as one of our isotope detectives. Boron has only two stable isotopes, ^{10}B and ^{11}B, which are normally found in the ratio of roughly 20 percent to 80 percent; the official standard for the $^{11}B/^{10}B$ ratio is 4.0437. The element is present in seawater at 4.5 ppm. It is incorporated into the carbonate shells of the foraminifera with a bias of between 1.0 and 2.6 percent toward the lighter isotope. The key new insight that Boron brings is that its ability to participate in carbonate formation is a sensitive function of how acidic the environment is, and that sensitivity provides another handle on the amount of atmospheric CO_2.

CO_2 is slightly acidic. Because roughly one third of all the CO_2 we have released into the atmosphere over the last two and a half centuries has dissolved in the ocean, the ocean is becoming more acidic through the following chemical reaction:

$$CO_2 + H_2O \rightarrow H_2CO_3 \rightarrow HCO_3^- + H^+ \rightarrow CO_3^{-2} + 2H^+$$

where the two Hydrogen ions are indicators of an acidic solution. Acidity is measured on the pH scale, where pure water has a neutral value of 7.00, and a strong acid such as sulfuric acid is 2.75. The pH scale is logarithmic (like the earthquake Richter scale), meaning that each integer change implies an increase in Hydrogen ions by a factor of 10. The pH of seawater in the mid-eighteenth century prior to the burning of fossil fuels was 8.179, whereas today it is 8.069. This apparently modest change corresponds to an increase in H^- ions of 29 percent, almost all from newly dissolved CO_2. This ocean acidification is already affecting coral and planktonic growth.

Boron is present in seawater in two forms: $B(OH)_3$ and $B(OH)_4^-$. The ratio between the two is strongly dependent on the acidity or pH value. At a pH of 7.5, the ratio of the neutral molecule to the ion is 12:1, whereas at a pH of 8.5, it's 1:1. The isotopic ratio is also different for the two molecules; at the current pH of 8.069, $B(OH)_3$ is 4.5 percent above the standard value, and $B(OH)_4^-$ is 1.75 percent above the standard (established for pH = 7.5).

Using an ocean sediment core from the northeast Atlantic, Marcus Gutjhar and colleagues derived a continuous set of isotopic measurements from 300,000 years before to 500,000 years after the PETM.[5] They found that the $^{13}C/^{12}C$ ratio falls precipitously by 0.34 percent simultaneously with a decrease in the $^{11}B/^{10}B$ ratio of 0.17 percent. These changes follow the upward change of +0.12 percent in the $^{18}O/^{16}O$ ratio that heralds the temperature rise by about 25,000 years (again, we're back to "sudden" on geological timescales). They find that the Boron ratio excursion implies a pH change of about 0.3, corresponding to a factor-of-2 change in the number of Hydrogen ions. Similar changes have been observed in cores from the South Atlantic and the equatorial Pacific, indicating a global phenomenon. This ocean acidification event allows the researchers to estimate the amount of new Carbon that must have shown up in the atmosphere: 10,000 gigatons, or roughly ten times the amount we have added from fossil fuel burning. This would produce an atmospheric concentration of CO_2 of over 2,000 ppm, five times today's value. It led to a sea-surface temperature rise of about 5°C (9°F). They infer that most of this emission occurred within 50,000 years of the onset of the acidification event.

While the total CO_2 emission and resultant atmospheric and ocean changes are much greater than in our current ongoing geochemical experiment, the *rate* of current acidification is much faster today. Their model suggests an initial CO_2 input rate of 1 gigaton of Carbon per year declining over the 50,000-year interval to less than 0.1 gigaton per year. Our current output amounts to 10 gigaton per year, ten to 100 times the rate during the PETM.

Many sources have been suggested for this massive input of Carbon to the atmosphere, including CO_2 from volcanoes; CO_2 and CH_4 from the melting permafrost; and CH_4 from methane clathrates, ice-like crystals of methane and water prevalent on the ocean floor. Because these sources have very different

$^{13}C/^{12}C$ ratios—+0.24 percent, −1.8 percent, and −5.6 percent, respectively—the observed change at the PETM allows us to infer which source is most likely. The estimated $^{13}C/^{12}C$ ratio of the new Carbon is −1.1 percent, which suggests a major contribution from vulcanism, with possible smaller contributions from the other sources. For example, an initial injection of volcanic CO_2 from the authors' preferred source, the North Atlantic Igneous Province,[6] could have warmed the Earth, leading to a feedback effect that contributed more CO_2 and methane from rotting plants in the melting permafrost. Such amplifying feedbacks are a matter of concern in the current era, with its rapid rise in the atmosphere's Carbon content.

ATMOSPHERIC CHANGES: END OF THE LAST ICE AGE

Twenty thousand years ago, there was a pile of ice almost a mile thick over where my office in Pupin Labs on Columbia's Manhattan campus stands today. This was a piece of the Laurentide Ice Sheet that stretched from the Canadian Arctic to as far south as New York and St. Louis and west to the Rockies (where it merged with the 1-million-square mile Cordilleran Ice Sheet that covered the mountains of western Canada as far south as Seattle). This massive quantity of ice had begun forming 75,000 years earlier and, along with large glaciers in Siberia and northern Europe and smaller ones in Patagonia and the Himalayas, contained enough water to lower global sea levels by over 400 feet. As discussed in chapter 11, the trigger for this event and those that preceded it was variations in the Sun's heating of the Earth caused by changes in Earth's orbital parameters, although strong feedback effects on the surface of the planet must have been involved to produce such dramatic changes.

Then the ice started to melt. The Earth's axis was moving toward a maximum tilt, the precession cycle was approaching the phase where the Earth is closest to the Sun in the Northern Hemisphere summer, and the orbit's ellipticity was approaching a local maximum. All three effects mean more intense sunlight in the far north; the optimal conditions were reached a little less than 12,000 years ago when the ice sheets began their terminal collapse, although the process began 6,000 years earlier.

It is easy to calculate the total change in solar energy reaching the northern ice sheets, and it is just too small to melt all that ice. As we saw in chapter 11, the amount of CO_2 in the atmosphere rises rapidly with rising temperature. The temperature in Greenland rose almost 9°C (16°F) over the next 5,000 years, while the atmospheric CO_2 concentration went from 180 ppm to 280 ppm over the same interval. All that extra CO_2 warmed the Earth further because of the greenhouse effect (chapter 11), while the newly uncovered land absorbed more solar energy than the reflective ice, which created a positive feedback loop that in turn led to ever-accelerating heating. But were these effects enough?

The answer is almost certainly no, which suggests additional feedback loops must be involved. A decade ago, G. H. Denton and collaborators published a comprehensive model for the collapse of the last ice age that attempted to take the complex interactions of ice, air, the oceans, and sunlight into account.[7] For the critical period between 20,000 and 10,000 years ago, they assembled an impressive collection of by now familiar isotopic, atomic, and molecular proxies: $^{18}O/^{16}O$ ratios from Greenland ice for temperature and from Chinese speleothems and Texan caves for precipitation, $^{2}H/^{1}H$ ratios for temperature and atmospheric CO_2 concentrations from Antarctic ice, and glycerol dibiphytanyl glycerol tetraether (GDGT) lipids for sea-surface temperature from deep-sea cores from the North Atlantic and off Australia. To these they added a novel proxy: the ratio of Protactinium-231 to Thorium-230 from a core off Bermuda as a measure of changes in the ocean's circulation pattern. Both ^{231}Pa and ^{230}Th are produced at a steady rate in seawater by the following decays:

$$^{235}U \rightarrow {}^{231}Th + \alpha \rightarrow {}^{231}Pa + e^- + \bar{v}_e \text{ and } {}^{234}U \rightarrow {}^{230}Th + \alpha;$$

$$(t_{1/2} = 7 \times 10^8 \text{ years})(\, t_{1/2} = 25 \text{ hours}) \quad (t_{1/2} = 2.5 \times 10^5 \text{ years})$$

They have a ratio in seawater of $^{231}Pa/^{230}Th = 0.093$. Both attach to particles in the water that are settling to the bottom, but Pa takes ten times longer to do so than does Th. Thus, the Pa is readily transported away by the circulation of the Atlantic's deep currents before settling, leading to a lower Pa/Th ratio in the sediment. If the deep current shuts down, the ratio will revert to the production ratio of 0.093.[8]

The researchers also included measurements of summer temperature in Patagonia from the extent of the glaciers there, deposits of debris on the ocean floor dropped by melting icebergs off the coast of Portugal, and opals on the sea floor off Australia. Opals are just basic sand (Silicon and Oxygen—SiO_2), with some water (H_2O) included (typically 6 to 10 percent water). Diatoms, single-celled algae, use the silicic acid (H_4SiO_4) in seawater to make tiny opal shells that accumulate, as with foraminifera shells, in the ocean sediment. The most prolific production occurs in the Southern Ocean, where lots of upwelling occurs and carries nutrients (particularly Silicon) to the surface, where it can be used by the algae. Thus, like the Pa/Th ratio, the opal content in deep-sea cores is a measure of oceanic circulation patterns.[9]

These ten proxies show a complicated pattern that leads to a consistent global explanation for the period from 20,000 to 10,000 years ago as the huge ice sheets started to collapse. In two intervals, between 18,000 and 14,500 years ago, and then in a shorter period between 12,800 and 11,500 year ago, temperature and precipitation actually *fell* in the Northern Hemisphere as large parts of the North Atlantic froze in the winter; in the Southern Hemisphere, however, CO_2 and temperature were both on the rise; the sea-surface temperature off Australia rose from 11°C to 16°C in the first period and from 16°C to nearly 20°C in the second, while the CO_2 rose from its glacial value of 185 ppm to roughly its preindustrial value (280 ± 15ppm) by the end of the second period. These same intervals also show dramatic shutdowns of the global ocean circulation. The shorter and more recent 1,300-year period is called the Younger Dryas (recently dated with remarkable precision using ^{14}C and dendrochronology to 12,807 ± 12 years before the present [BP])[10] and shows a sudden 0.7 percent decline in the temperature proxy $^{18}O/^{16}O$ to start the period and a similar rise when it ended, with both transitions occurring in less than a few decades (a sobering thought—such a rapid change in climate today would have devastating consequences).

The researchers interpret all these data within the context of a model driven by ocean circulation changes. As the huge northern ice sheets start to melt, they release large amounts of fresh water into the ocean between Iceland and Norway where, today, the warm, salty Gulf Stream finally cools enough for the water to sink to the bottom of the ocean, which in turn drives the worldwide ocean circulation pattern. The sudden addition of fresh water from the European and Laurentide ice sheets made the Gulf Stream water less salty and therefore less dense, so it failed to sink, shutting down ocean circulation. This stopped the warm water from moving toward the North Atlantic, which then froze in the winter, plunging the Northern Hemisphere into a cold snap. In the Southern Ocean, however, melting of the sea ice around Antarctica was well underway, and the warming continued apace. Complex feedback loops, including shifts in atmospheric circulation patterns, end up transporting heat north to end the cold snaps and promoting further melting of the ice.

Consistent with this picture is the evidence for two rapid periods of sea-level rise whose only plausible source is the melting of vast quantities of glacial ice. The two meltwater pulses described above are dated at 14,700 to 13,500 years ago and 11,500 to 11,200 years ago, each following closely the two periods of cold in the north. In the first period, sea level rose at an average rate of up to 2 feet per decade, roughly fifteen times the current rate.

Even with all these ingredients, it is not clear we have the complete recipe that frees the Earth from a global ice age. Fortunately, new ingredients keep popping up (so to speak). Craters up to 1 km across have been discovered on the seafloor in the Barents Sea north of Norway. They are formed when methane (CH_4), slowly diffusing up from deep (more than 1 km) oil and/or gas deposits, gets trapped in icy clathrates a few hundred meters below the surface. When the sea was frozen into the glacial ice, the pressure kept the methane in place. But when the ice retreated, the pressure was released and the methane burst through the surface, shooting up through the water and into the atmosphere. The amount of methane contributed by this process worldwide is unknown because the seafloor has not been mapped in the requisite detail to see all these small craters. But the ice core measurements show the atmospheric methane content roughly doubled from the glacial (350 ppm) to the interglacial (700 ppm) periods, and this could well be one of the sources, which add to greenhouse warming.[11]

THE FUTURE

At present, our atmosphere's composition is relatively stable. The obvious exceptions are the anthropogenic changes in the rare but crucial gases of carbon dioxide (CO_2), nitrous oxide (N_2O), methane (CH_4), and the chlorofluorocarbons that enhance the greenhouse effect and will warm the planet by a few degrees before the end of the century unless immediate actions are taken to mitigate these changes. The lightest gases that can reach escape velocity—Hydrogen and Helium—are still escaping as soon as they are produced, hydrogen as a consequence of the UV dissociation of water in the stratosphere (see below) and helium from the alpha decay of radioactive elements in the crust and oceans (recall that alpha particles are just Helium nuclei). The Hydrogen loss rate is about 93,000 tons a year, with Helium departures measuring 1,600 tons per year. These may sound like big numbers but, given the 150 billion trillion tons of Hydrogen in the oceans, it is a trivial effect . . . at least for now.[12]

Eventually, the Sun will swell to become a red giant, gobbling up Mercury and Venus and scorching the Earth; "eventually" in this case means in another 7 billion years. However, we won't have to wait that long to see yet another whole-sale transformation of our planet. Because of the fusion reactions in its core and the concomitant changes in its internal structure (chapter 16), the Sun is slowly brightening. In roughly a billion years, it will be about 10 percent brighter than it is today. This may seem like a modest change, but it is ten times greater than the heating induced by the greenhouse gases we have added to the atmosphere over the span of 200 years. The Earth's warmer surface will lead to more evaporation of H_2O from the oceans, but it is the effect this heating will have 10 to 20 km above the Earth's surface that's most important.

The troposphere is the layer of air from the ground up to an altitude of roughly 20 km at the equator and 10 km at mid-latitudes. This is the layer within which weather takes place. Throughout this layer, the temperature falls steadily from an average of 16°C at the surface to a chilly −50°C at the top. (That's −60°F—jets fly at these altitudes in the troposphere, so check the outside temperature on the flight monitor on your next trip.) Above this layer is the stratosphere, extending upward to 50 to 60 km. Throughout this layer, the temperature rises again, reaching roughly 0°C at the top. In the lower stratosphere is a layer of ozone (O_3) molecules that block most of the Sun's UV light from reaching the lower atmosphere and the planet's surface.[13]

The temperature minimum at the boundary between the troposphere and stratosphere means the air there is very dry. First, cold air can hold less water vapor than warm air, and second, most of the water vapor condenses to a liquid as it rises and cools in the upper troposphere, falling as rain or snow before even reaching the boundary. Thus, very little water ascends to the middle stratosphere. This is critical because it is only there, above the protective ozone layer, that the Sun's UV light can blast apart the H_2O molecules and allow the H_2 to escape.

As the Sun brightens and the boundary layer warms, however, the leakage of Hydrogen will go from the trickle it is today to a torrent a billion years hence. Within another billion years, most of the oceans will have evaporated, and the Earth will be a hot, dry, ice-free place. A billion or two years after that, all water will be gone and the Earth will resemble Venus, a planet where this process went to its conclusion within the first billion years of the solar system's existence, leaving that planet with a CO_2-dominated atmosphere and a surface temperature of 460°C (almost 900°F). Life as we know it will be extinguished. The good news is that we have lots of time (10,000 times longer than the current reign of *Homo sapiens*) to figure out what to do.

CHAPTER 15

Our Sun's Birthday

The Solar System in Formation

My paternal grandmother was from Belarus and had no idea when her birthday was. This was not a consequence of any loss of mental faculties—she was sharp as a tack up to her death at age ninety-nine. It's just that it wasn't very important if you were a late nineteenth-century peasant in Czarist Russia to remember the day you were born.

I suppose the same could be said about the solar system—figuring out its exact birthday isn't a burning issue of great import. But it would satisfy an idle curiosity. And such a history might reveal the processes by which the planets came to be arranged the way they are, while also extending the reach of our atomic historians beyond the lifetime of Earth itself. In fact, we can even gain some insight to what was happening in our neighborhood before the Sun formed.

FORMATION SCENARIO

Our galaxy, the Milky Way, is a collection of more than 200 billion stars held together by their mutual gravitational attraction. It began as a tiny fluctuation of excess density in an initially remarkably uniform, purely gaseous universe. Gradually, over 1 billion years or more, this dollop of extra mass accreted surrounding material, slowly growing in mass and enhancing its gravitational attraction for ever more distant material. As it contracted, an initial random rotation picked up speed, much as a figure skater draws in her arms in order to spin more rapidly.[1] This increasing spin caused much of the normal matter to collapse into a relatively thin disk, producing the pinwheel-shaped galaxy we see today only a few thousand light years thick but 100,000 light years across.

While initially composed primarily of gas,[2] the process by which the Milky Way separated from its surroundings—accreting more and more matter onto a tiny random density peak—is reproduced in microcosm as stars start to form from fluctuations in the gaseous disk. By now, 13 billion years on, roughly 90 percent of the galaxy's gas is to be found in stars, with only 10 percent remaining to form future generations of stars and planets.

A population of more than 200 billion stars may make the galaxy sound like a crowded place, but that's hardly the case. If we represent the Sun (almost a million miles in diameter) by an orange in New York, the Earth is about the size of a grain of sand 15 feet away. The most distant planet, Neptune, is a small pea two city blocks away, and the next star, similar in size and temperature to the Sun so appropriately represented by a second orange—is in Minneapolis. Stars make up only a 100 billionth of a trillionth (approximately 10^{-23}) of the volume of the galaxy. Most of the Milky Way, then, is what we call interstellar space.

This space is not completely empty—there's still that 10 percent of the original gas that has not yet formed stars. Most of this, like most of the universe, is Hydrogen and Helium, but it is far from uniformly distributed. In the majority of interstellar space, the gas particles are very diffuse, with only 1 atom per 100 to 1,000 cubic centimeters (i.e., just a single atom of Hydrogen in a one-quart container, 10,000 times less dense than the best vacuum we can create in the laboratory). The temperatures in these regions range from a few hundred thousand to a few million Kelvins, so most of these atoms are ionized. The high temperatures and low densities are created and maintained by the stellar explosions we'll explore in chapter 16.

In a few percent of the interstellar medium (as we call the material between the stars), the gas has managed to cool and condense to much higher densities, from 1 to 100 atoms per cubic centimeter at temperatures of 100 to 200 K. In a few percent of this few percent, even denser regions are encountered, with thousand to tens of millions of atoms per cubic centimeter and temperatures as low as 10 K. In these dense, cold regions, molecules can form, and microscopic grains of dust made primarily of Carbon, Silicon, and Iron are found. These cold clouds can contain material from hundreds to a million times the Sun's mass. It is here that new solar systems are born.

Again, a small random fluctuation in density inside one of these cold dark clouds starts to grow by accreting surrounding material. A stellar core forms at the center surrounded by a spinning disk of material stretching out 100 billion km and more. As the collapsing cloudlet continues to shrink and detaches from the surrounding material, it becomes a protostellar nebula, precursor to a new solar system. Typically, greater than 99 percent of the matter ends in the central star (99.86 percent in the case of our solar system), but the surrounding disk still contains approximately 3 trillion trillion tons (3×10^{27} kg) of matter that constitutes the raw material from which planets, moons, asteroids, and comets are formed.

The details of this formation process are beyond the scope of this book (and, in part at least, beyond our current knowledge), but it is clearly a bottom-up process: dust grains stick together to make bigger grains, the bigger grains stick together to make tiny pebbles, pebbles come together to make rocks, and so on. Near the Sun, it is too warm for ices (of water, carbon dioxide, ammonia, etc.) to form, and many of these gases, along with the dominant Hydrogen and Helium, are evaporated from the inner solar system (thus explaining the lack of these gases on Mercury, Venus, Earth, and Mars). Farther out, however, between the current orbits of Mars and Jupiter, it is cold enough for these ices to freeze, and they add to the accumulating planetesimals, building planetary cores quickly enough that the Hydrogen and Helium gases remain bound, accounting for the gas (Jupiter and Saturn) and ice (Uranus and Neptune) giant planets of the outer solar system.

The vast majority of the material in the original disk gets swept up into one of the planets, but gravitational vacuum cleaners aren't perfect, and a fraction of 1 percent of the matter never finds a planetary home. These fragments (rocky and metallic asteroids in the inner solar system, icy Pluto-like objects and comets farther from the Sun) thus retain a record of what the solar system was like in its very earliest days. When one of these fragments falls to Earth as a meteorite, we can hold it in our hands and look upon a piece of the primordial solar system. Interrogating its atoms allows us to determine its age and probe the prehistory of our home.

THE EARLIEST MATERIAL

As noted in chapter 13, the dynamic process of plate tectonics constantly reshapes the Earth's surface, leaving no rocks accessible from the planet's earliest days. Even if it were not for this constant refinishing of Earth, however, the modifications that rocks would undergo slamming into an initially molten Earth would erase some of the clues to their origin. The meteorites, then, provide a more useful, pristine sample to interrogate in constructing our origin story.

Several different types of meteorites fall to Earth, their composition and structure reflecting their different origins. Iron meteorites which are at least 95 percent Iron, Nickel, and Cobalt come from large bodies (large asteroids or protoplanets that didn't quite make it to planethood before being broken up in a collision); these bodies had sufficient mass to become molten, and the heavy metals sank to the center (this segregation of the elements by mass is called differentiation and is responsible for the Earth's rocky crust and Iron core). Stony meteorites are, as their names suggest, like rocks, composed mostly of Silicon- and Aluminum-based minerals; some of these are pure stones and thus are likely unprocessed from the time of their original formation, while others are mixed with some metal and may have come from differentiated asteroids. Small numbers of meteorites actually come from the Moon and from Mars; when a large meteor collides with one of those bodies, it blasts some shards of the crust into space, where they wander until colliding with Earth.

The most primitive, least processed meteorites making up just under 5 percent of the total that reach Earth are the carbonaceous chondrites. Some of these contain a significant amount of water (from a few percent up to more than 20 percent), implying they cannot have been heated to high temperatures during their formation or the water would have evaporated into space. As discussed in chapter 13, many of these are also rich in organic compounds, including amino acids. Of most interest for our current project, however, is that such bodies also contain chondrules and Calcium- and Aluminum-rich Inclusions (CAIs), the earliest solid compounds to form in the young solar nebula.

Chondrules are small globs of minerals ranging in size from a hundredth of a millimeter to 1 cm across. They are primarily composed of Silicon and Oxygen with admixtures of varying amounts of Aluminum, Magnesium, Potassium, Calcium, Phosphorus, Chromium, and other such elements between number 11 (Sodium) and number 26 (Iron) in the Periodic Table (we'll see why just this range in chapter 16). They are believed to be the remnants of dust grains in the solar nebula that were quickly (in minutes) melted when heated to a temperature of approximately 1,000 K. The source of this flash heating is uncertain, but it could include solar flares, shock waves in the rotating disk of material in which they were embedded, collisions with larger bodies, and so on. Following the heating, they quickly recondense to solid form and are accreted by large bodies, where they are incorporated into the rocky matrix of, say, an asteroid. Some chondrites are glassy (the molecules are jumbled together in an irregular pattern), while others are crystalline (where the molecules form a highly regular lattice).

CAIs are found only in carbonaceous chondrite meteorites. They are similar to chondrules, but they are apparently formed at a higher temperature (more than 1,300 K); whether they are systematically even older than chondrules has been a matter of debate, although recent evidence, discussed below, suggests similar ages dating from the first approximately 1 million years of the solar system. They consist of various minerals such as anorthite ($CaSi_2Al_2O_8$), perovskite ($CaTiO_3$), and forsterite (Mg_2SiO_4), among many others.

THE BIRTH DATE

To date something as old as the solar system requires radioactive isotopes with long half-lives (1 billion years or more). In addition, however, we cannot use the simple accumulation clock in which one radioactive isotope decays to a stable daughter isotope unless we know how much of the daughter was there to start with; the lack of geologists standing by as the solar system formed puts us in a bind. But a clever technique using isochrones (literally "equal times") gets us out of this dilemma. To illustrate, I will show the Rubidium-Strontium isochrone construction, which has been used to date chondrules as well as many rocks from the Earth and Moon (figure 15.1). The results from Lead-Lead dating, a slightly complicated twist on this basic method, yields the most precise solar system birth dates and will follow this initial explication.

Rubidum-87 (^{87}Rb) undergoes a standard beta decay to Strontium-87 (^{87}Sr) by spitting out an electron and an antineutrino; the half-life is 48.8 billion years. The problem is, ^{87}Sr is already present in an unknown quantity in the sample before the radiogenic atoms of this isotope start appearing after the Rb decay. To solve for this unknown, we also measure the amount of ^{86}Sr, a stable, nonradiogenic isotope, in the sample. The algebra necessary to solve uniquely for the age of the sample is spelled out in the notes.[3]

Given the long half-life of Rb, less that 7 percent has decayed since the beginning of the solar system. But ^{238}Uranium has a half-life almost perfectly matched to the time we are trying to measure, 4.5 billion years. For the meteoritic

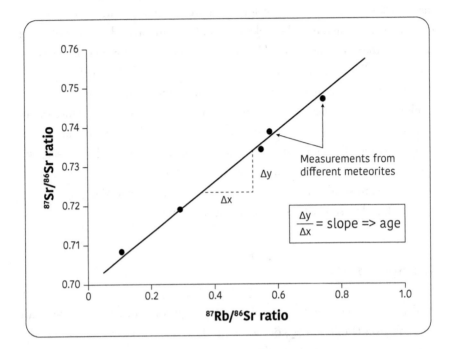

15.1 Isochrone curve for the Rubidium-Strontium dating technique. The slope of the line allows one to derive unambiguously the age of the mineral in question, in this case, from meteorites formed at the time of the solar system's birth (see the text and chapter note 3.)

chondrules and CAIs, the most accurate dates have been derived from a variation on the Uranium-Lead dating technique (described in chapter 9) called Lead-Lead dating.

Recall that ^{238}U decays to ^{206}Pb, and ^{235}U decays to ^{207}Pb. In addition, there is a nonradiogenic isotope of Lead, ^{204}Pb. By plotting $^{207}Pb/^{206}Pb$ versus $^{204}Pb/^{206}Pb$ (and taking into account very slight variations in the initial $^{235}U/^{238}U$ ratios in the solar nebula), one can construct isochrones to obtain an age for a CAI from the Efremovka meteorite of 4,567.35 ± 0.28 million years ago (Mya), while chondrules from other meteorites range from 4,567.32 to 4,564.71 Mya.[4] Besides being an easy number to remember (4,5,6,7 Mya), this suggests that CAIs and the earliest chondrules formed at the same time, although the latter perhaps had an extended period of formation over 2 or 3 million years. The accuracy of this date is noteworthy: an uncertainty of 0.28 Myr out of a total of 4,567 Myr is equivalent to your being able to tell my age to within 1.5 days, a feat not possible with or without radioactive dating!

BEFORE THE BEGINNING

In addition to the long-lived radioactive isotopes present in the early solar system that help us determine its age, there are also the decay products of much shorter-lived isotopes. One whose role was clearly important (but whose origin is controversial) is Magnesium (^{26}Mg), which is produced when Aluminum (^{26}Al) decays with a half-life of only 717,000 years. This inverse beta decay, which leads the nucleus to step down one spot in the Periodic Table through the emission of a positron, is very energetic, yielding 4 million electron volts (MeV) per decay. In addition, the Magnesium is left in an excited state and subsequently emits a gamma ray photon with an energy of 1.808 MeV.

The ^{26}Al isotope is produced in massive stars as they undergo their various stages of nuclear fusion (see chapter 16). In the rare, very massive stars (more than thirty or forty times the mass of the Sun), ^{26}Al can be dredged up from the interior and blown into space by the strong stellar winds characteristic of such stars in the late stages of their lives. For all stars more than 8.5 to 10 times the mass of the Sun that end their lives by exploding, their ^{26}Al is distributed throughout space with the other elements that star has produced. We know these processes are ongoing in the Milky Way Galaxy today because our gamma ray telescopes have detected the 1.8 MeV photons from the ^{26}Mg decay, allowing us to infer that at any one time, there are about two solar masses of ^{26}Al scattered throughout interstellar space.

Given the huge volume of our galaxy, if this isotope were uniformly distributed, any given location, such as our protosolar cloud, would have an immeasurably small amount of ^{26}Al (less than 50 kg in the whole nebula). But because the ^{26}Al half-life is short on cosmological timescales, it doesn't uniformly blanket the whole galaxy, but it is concentrated in regions where stars are forming.

The only isotope of Aluminum that lives more than 10 minutes besides ^{26}Al is the one stable form, ^{27}Al. Thus, the ratio of ^{26}Al/^{27}Al in the early solar system that we can infer from the current abundances of ^{26}Mg and ^{27}Al is ^{26}Al/^{27}Al $= 5 \times 10^{-5}$ (although estimates a factor of 2 lower have also been deduced, suggesting the distribution of the radioactive species may not have been uniform throughout the disk). This number is interesting because it has the potential to solve another mystery of the early solar system—the differentiation of the asteroids.

A planet like the Earth accumulates so much energy from the massive number of planetesimals raining in on it during formation that it gets very hot—hot enough to melt rocks and allow the heavy elements to collect in the core while the lighter ones float to the surface—this density sorting is the differentiation mentioned in chapter 12. But for smaller bodies like asteroids only tens to hundreds of kilometers in diameter, the gravitational energy released during accretion is insufficient to melt them, and we should expect them to resemble a rubble pile composed of the bits and pieces from which they were made rather than a smoothly differentiated body. Given that we find both purely metal and purely stony meteorites, however, it is clear that even these small bodies somehow, shortly after their formation, melted and differentiated too. A sufficient source of energy is radioactive heating.

Even at an average abundance of 2.5×10^{-5} ^{26}Al atoms per ^{27}Al atom (25 parts per million [ppm]), radioactive decays will deposit 3,000 joules of energy per gram of material, far exceeding the gravitational energy from accretion and more than enough to melt an asteroid and allow differentiation to proceed. The origin of this excess of ^{26}Al in the early solar system remains controversial. The original idea was that a massive star, born from the same cloud as the Sun, exploded nearby, seeding the gas with radioactive isotopes and possibly also triggering our cloudlet to begin its collapse. We see this process underway today in distant interstellar clouds undergoing star formation. This scenario also has the attractive side benefit that it would have produced a nearby neutron star that could have triggered the preference for left-handed amino acids we find throughout the solar system (see chapter 13). Serious objections to this picture have been raised, however. Besides the a priori implausibility of having a supernova go off right next door when there are only a few per century in the whole galaxy, most models that yield enough ^{26}Al produce too much Manganese-23 and Iron-60 to explain the low abundances of their daughter nuclei in the primordial solar system material.[5]

Alternative hypotheses include the delivery of the Aluminum on interstellar grains created by a more distant supernova, or the winds from very massive stars mentioned above (although it should be noted that such massive stars are also very rare, so the implausibility argument again rears its head). The idea that high-energy particles from the young Sun might have produced the radioactive isotopes directly in the protoplanetary disk is hard to make consistent with what we know about the activity of young stars and the abundances of other short-lived isotopes. Our ability to measure ever-smaller quantities of rare isotopes and their daughter products may clarify the situation in the near future.

As noted earlier, the chondrules and CAIs are the earliest solid material we have from the solar system's birth, created within a million years or so of the formation of our protoplanetary disk. It probably took roughly 10 to 20 million years for the Earth to sweep up all the material in the vicinity of its orbit and cool enough to have a solid crust. Shortly thereafter, a large protoplanetary object (possibly nearly the size of Mars) collided with the Earth and ejected enough material to create the Moon. After a period of relative tranquility, between 4.1 and 3.8 billion years ago, we entered what is called the Late Heavy Bombardment Era, during which massive bodies continued to crash into the Earth, Moon, and other planets. But, as noted in chapter 12, zircons from up to 4.4 billion years ago indicate there was already liquid water on the Earth's surface at that time, and by the end of the bombardment era, the first signs of life appear.

Our atomic historians have allowed us to reconstruct history back to the first years of the solar system's existence. Over the past two decades, our discovery of thousands of extrasolar planets orbiting their own stars marks this history as less than extraordinary. Nonetheless, it is *our* history, and dating its beginning so precisely is gratifying. Now it is time to turn to the history of the historians themselves to see how the particular arrangements of the leptons and quarks that make up our world, so helpful throughout, came to be.

CHAPTER 16

Stardust Creation

Building the Building Blocks

Ll historians have parents. This is equally true for the atomic historians that have been our guides throughout this book. Indeed, the ninety-one members of our family of elements numbered 4 to 94 all have the same parents: the primordial elements Hydrogen, Helium, and Lithium with which the universe began, although additional amounts of the latter two are still being produced today. As we will see in chapter 17, these parents had antecedents themselves, although those grandparents were only around for a few minutes before spawning the three lightest nuclei that comprise the raw material of the cosmos.

The fact that, after more than 13 billion years, approximately 99 percent of the matter in the universe is still in the form of Hydrogen and Helium (chapter 13) suggests creating all the other elements isn't easy. This should come as no surprise if you recall how an atomic nucleus is put together—a bunch of positively charged protons that fiercely repel each other get squeezed into a space less that one trillionth of a centimeter across before the attractive strong force kicks in, and even this force isn't sufficient without the padding and added glue of a bunch of neutrons crammed into the same tight space.

Constructing atomic nuclei thus requires extreme conditions. To have a chance of locking together, the constituents of nuclei must be tightly packed and moving at very high speeds; this translates to a requirement of high densities and high temperatures. The places where such conditions are met include accretion disks around black holes, in the cataclysmic collisions of two neutron stars, and in the cores of regular stars. Because the first twenty-six elements (up to Iron) make up more than 99.9999 percent of the atoms in the universe, and most of these are cooked up in stellar cores, we'll start our atomic genealogy there.

STELLAR ELEMENT FACTORIES: START WITH HELIUM

The Sun appears to be the same size in the sky every day, as does the Moon.[1] The reason for these two constancies, however, is different. The Moon is a solid body, held in a spherical shape by gravity and held up against collapse by electromagnetic interactions among the atoms and molecules of which it is composed. Its central density is only about twice its average density of 3.35 g/cm^3.[2] The Sun, on the other hand, is not a solid but a plasma in which the electrons are liberated from their nuclei. This plasma state (chapter 3) leads to a very different density and temperature structure from that of a solid body such as the Earth or Moon. At the photosphere, the layer of the Sun from which visible light is emitted (thus, the layer we see), the density is about eight times that of the air on Earth (just 1 percent the density of water)—in other words, you could walk through it with only minor effort (were it not quite so hot). In the Sun's core, however, the density rises to 150 times the density of water, seven times denser than the densest substance on Earth (the metal Osmium, element number 76) or 15,000 times greater than the density at the Sun's surface. The temperature undergoes a similarly dramatic rise, from 5,780 K on the surface to 15,700,000 K at the center.

The Sun is thus far from a solid object—it is in a constant dynamic tension between gravity, which is trying to shrink it, and the thermal pressure inside, which is pushing outward to expand it. Because it is radiating at the rate of 380 trillion trillion watts, it would soon cool down, reducing the interior pressure and initiating shrinkage, were it not for the nuclear furnace in the core that is constantly resupplying the lost energy. Our constant Sun, then, is held in a delicate balance by two of the fundamental forces of nature: gravity pulling inward and nuclear fusion (courtesy of the strong force) pushing outward.

Like the rest of the universe, the Sun is mostly composed of Hydrogen, so that is the primary fuel for the nuclear reactor. The process proceeds in three steps (see figure 16.1). First, two protons—bare Hydrogen nuclei—need to stick together. This is challenging because their positive charges repel each other, and the closer together they get, the stronger the repulsion. Recall from chapter 3 that they must get within 10^{-14} meters of each other in order for the strong nuclear force to take over and snap them together. This doesn't happen very often. Indeed, the odds of it happening for any given proton are such that it should occur about once in 10 billion years. Fortunately, there are lots of protons in the Sun, and so,

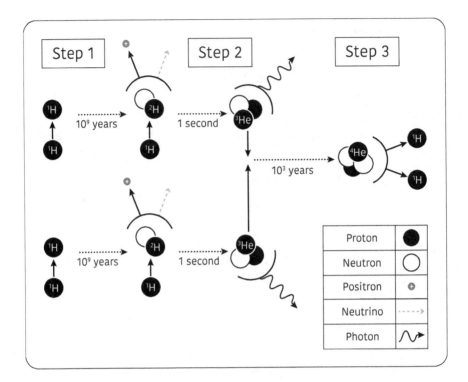

16.1 The three-step proton-proton chain reaction that creates a ⁴He nucleus from four protons. This sequence of reactions powers the Sun and most other stars. See the text for details.

despite these daunting odds, it happens about 92 trillion trillion trillion (yes, three of them: 92×10^{36}) times per second:

$$^1H + {}^1H \rightarrow {}^2H + e^+ + v_c$$

The next step in the process is speedy because Deuterium (^2H) is highly reactive, and within a few seconds, it connects with another proton to make the light isotope of Helium:

$$^2H + {}^1H \rightarrow {}^3He + \gamma$$

The final step requires two of these Helium nuclei to fuse in the following reaction:

$$^3He + {}^3He \rightarrow {}^4He + {}^1H + {}^1H$$

yielding the dominant isotope of Helium and releasing two protons available to continue their participation in the fusion process. The net result is that four protons end up as two protons and two neutrons bound together as a very stable Helium nucleus.

And, of course, energy is released—lots of energy—in an amount equal to the binding energy of a ^{4}He nucleus. In step 1, the positron produced quickly finds an electron with which it will annihilate, releasing 1.022 million electron volts (MeV) in two photons; the neutrino also carries away some energy, meaning this step (which must occur twice to make two deuterium nuclei) produces 2.884 MeV in total. ^{3}He is more stable than Deuterium, so it releases even more binding energy when it snaps together (recall how this works from chapter 3): 5.49 MeV, times two again to make two ^{3}He nuclei so 10.98 MeV in total. The final step, which leads to the stable (therefore tightly bound) ^{4}He, releases 12.859 MeV. The net yield from the completed three-step process is 26.73 MeV. Taking into account the number of times per second this happens (given above), we get a total rate of solar energy output (the Sun's "luminosity") of 3.92×10^{26} J/s or 3.92×10^{26} watts (a lot of light bulbs). About 2.5 percent of that energy comes out with the neutrinos, so the electromagnetic output (light, X-rays, ultraviolet [UV] photons, etc.) is 3.828×10^{26} watts.

Where does the energy come from? As described in chapter 3, it is transformed from mass into energy through $E = mc^2$—the resulting Helium nucleus weighs less than the sum of its constituents because mass is lost to create the nuclear binding energy, and all that fusion energy is released. Each second, the Sun converts about 600 million tons of Hydrogen to 595.75 million tons of Helium, and 4.25 million tons of matter disappears from the universe, reemerging as light. For reference, 4.25 million tons of matter is enough to fill a coal train stretching from New York City to Montreal, Canada (350 miles). Snap your fingers once per second and with each snap, one of those trains disappears.

In reality, the process is somewhat more complicated because several other reactions can occur (e.g., a ^{3}He can find a ^{4}He and temporarily make ^{7}Be—see box 16.1), but the basic proton-proton (p-p) reaction produces 82 percent of the Sun's energy, and the alternative chains produce the rest. These reactions, however, are very sensitive to temperature and pressure in the stellar core, as well as to the presence of the nuclei of other elements, like Carbon, Nitrogen, Oxygen, and so on. In stars 30 percent more massive than the Sun, the core temperature

---

BOX 16.1 BEYOND THE PROTON-PROTON (P-P) CHAIN

The p-p chain dominates the Sun's energy production (82 percent), but several additional nuclear fusion reactions contribute about 18 percent of the total energy, and in more massive stars, they become more important. Several of these reactions are listed here:

p-p branch II: This contributes about 16 percent of the Sun's total energy.

$^{3}He + {}^{4}He \rightarrow {}^{7}Be + \gamma$

$^{7}Be + e^{-} \rightarrow {}^{7}Li + \nu_{e}$

$^{7}Li + {}^{1}H \rightarrow {}^{4}He + {}^{4}He$

p-p branch III: This contributes about 0.01 percent of the Sun's total energy.

$^{3}He + {}^{4}He \rightarrow {}^{7}Be + \gamma$

$^{7}Be + {}^{1}H \rightarrow {}^{8}B + \gamma$

$^{8}Be \rightarrow {}^{8}Be + e^{+} + \nu_{e}$ or $^{8}Be \rightarrow {}^{4}He + {}^{4}He$

Carbon-Nitrogen-Oxygen (CNO) cycle (see figure 16.2): This contributes about 2 percent of the Sun's total energy.

$^{12}C + {}^{1}H \rightarrow {}^{13}N + \gamma$

$^{13}N \rightarrow {}^{13}C + e^{+} + \nu_{e}$ (half-life 10 minutes)

$^{13}C + {}^{1}H \rightarrow {}^{14}N + \gamma$

$^{14}N + {}^{1}H \rightarrow {}^{15}O + \gamma$

$^{15}O \rightarrow {}^{15}N + e^{+} + \nu_{e}$ (half-life 2 minutes)

$^{15}N + {}^{1}H \rightarrow {}^{12}C + {}^{4}He$

---

is just enough higher to allow a different fusion cycle to become dominant (see box 16.1 and figure 16.2 for an illustration of the CNO cycle). But whatever their mass and temperature, stars are shining throughout most of their lives by transforming Hydrogen to Helium. In the 13.8-billion-year history of the universe, the 100 billion trillion stars have managed to convert about 2 percent of the Hydrogen to Helium, so there is still a long way to go before we run out of nuclear fuel.

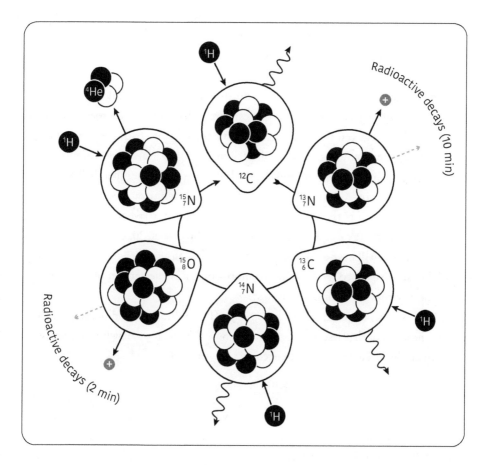

16.2 The six-step CNO cycle that contributes only about 2 percent of the Sun's energy but becomes the dominant energy source for stars just 30 percent more massive than the Sun. Note the importance of two short-lived radioactive beta decays. The net result, as in the proton-proton chain, is that four protons (^{1}H) are turned into one Helium nucleus (^{4}He), but the Nitrogen, Carbon, and Oxygen intermediaries give off three photons, two neutrinos, and two positrons (which in turn annihilate with electrons and produce four more photons).

STELLAR ELEMENT FACTORIES: WHAT COMES NEXT?

Hydrogen-to-Helium fusion is a one-way street. In principle, it is possible to reverse the process—to deconstruct a Helium nucleus into its constituent particles (a couple of protons and a couple of free neutrons that will, within 15 minutes or so, decay back into protons with the emission of an electron and an antineutrino), but this process is rare in the cores of stars—neither the nuclear collisions nor the photons have enough energy to exceed the 26.7 MeV binding energy of the Helium nucleus and break it apart.

As a result, the stellar core eventually runs out of fuel—when all the ^{1}H has been converted to ^{4}He, the nuclear reactor turns off. This disrupts the dynamic balance between gravitational and nuclear forces that the star has spent its lifetime maintaining, and the core of the star starts to shrink. As it does so, a layer of Hydrogen surrounding the now all-Helium core gets heated and compressed to the point where it can fuse, and the nuclear reactor reignites in a shell around the shrinking core. All this fresh fuel, and the higher temperatures and densities it encounters, boosts the reaction rate, swelling the outer layers of the star and increasing its energy output first by a factor of three, then 30, then 300. By this point, the Sun will have gobbled up Mercury and Venus, and its outermost layers will be far too close to the Earth for comfort—only a burned cinder of our planet will bear witness to what happens next.

By this time, the core is both hot enough (100 million K)[3] and dense enough that the ashes of the first reaction, ^{4}He, become the fuel for a new fusion process in which three ^{4}He nuclei fuse to become Carbon (^{12}C):

$$^4\text{He} + {}^4\text{He} + {}^4\text{He} \rightarrow {}^{12}\text{C} + \gamma$$

Because Carbon has a higher binding energy than Helium, 7.725 MeV is released in each such fusion. The ignition of this reaction briefly drives the luminosity to 2,000 times its current value (known as the Helium flash), but the star soon settles into a new equilibrium in which He → C in the core and H → He in a surrounding shell, a state that remains stable for about 150 million years (note that this is only a little more than 1 percent of the star's lifetime—the end is nigh).

Eventually, the center of the star becomes pure Carbon and, with the core reactor turned off, it again starts to shrink. This draws more virgin Hydrogen and Helium into density and temperature zones where they can ignite, and the star's luminosity skyrockets again to several thousand times its original value. But this arrangement—an inert Carbon core, a He → C shell around it, and a H → He-burning shell around that—is inherently unstable. As the reactions turn on, they inflate the shells, thus lowering both the temperature and density below the critical threshold, and the reactions turn off. This in turn causes the shells to shrink, reboosting the temperature and density until the fusion threshold is crossed and the reactions turn back on. The star starts to pulsate, and its extended outer layers get blown off into space.

This on-off cycle repeats on timescales of 100,000 years or so as the star loses up to 40 percent of its mass and the core shrinks ever smaller. Eventually, the game is up, and what's left of the star collapses to an object the size of the Earth (approximately 1 percent of the original diameter of the Sun). At this point the electrons (which we have largely ignored but which must be there to balance the positive charges of the nuclei) get squeezed so closely together that the rules of quantum mechanics intervene and forbid them from getting any closer. Recall from chapter 4 that no two electrons can have the same energy, angular momentum, and spin, leading to the energy shells and subshells of the Periodic Table. Here, the whole stellar remnant turns into what is effectively a giant macroscopic atom, with the electrons filling trillions and trillions of energy levels. The resulting force, called electron degeneracy pressure, is sufficient to balance the inward pull of gravity for stellar remnants up to 1.4 times the mass of the Sun.

In this configuration, the star is called a white dwarf. Its density is roughly 1 ton per teaspoonful—imagine a sports utility vehicle (SUV) compressed to the size of a sugar cube. The only source of energy it has left is the heat it retained from the peripatetic nuclear reactions of its death throes—it becomes like an ember in a fireplace, slowly radiating its heat energy into space, initially at a temperature of 100,000 K or more when it is first revealed by its cast-off outer layers, and gradually lower and lower temperatures as the heats radiates away. Its mass has effectively been taken out of the cycle of stellar birth and death in the galaxy. While a bit of the Carbon it produced can be dredged up during its final pulsations and expelled into space with the star's original outer layers, it has contributed very little to making the universe an atomically richer place.

The majority of stars in the Milky Way have lower initial masses than the Sun, and all follow a very similar life cycle. It seems counterintuitive, but the lower the mass of a star, the longer it takes to evolve;[4] even the universe's first generation of stars that began life with less than 80 percent the mass of the Sun are still in their Hydrogen-burning phases. If all stars followed this playbook, we wouldn't be here to discuss it because the primordial Hydrogen, Helium, and Lithium would have been changed only by the addition of a tiny fraction of Carbon, and the Galaxy would be littered with white dwarf corpses. Fortunately for us, more massive stars live more dramatic lives.

STELLAR ELEMENT FACTORIES: THE CONTRIBUTION
OF MASSIVE STARS

For stars that begin life with a mass several times greater than that of the Sun, Carbon production is not the end of the line. The extra overlying mass raises the density and temperature enough that some of the Carbon at the outer edge of the core can react with the Hydrogen in the H → He shell and produce Nitrogen via the first three steps of the CNO cycle (see box 16.1). In addition, if and when the temperature gets high enough, Helium nuclei can fuse with Carbon to make Oxygen, Neon, and Magnesium (the even numbered elements 8, 10, and 12, adding two protons and two neutrons at each step: $^{12}C + n \times {}^{4}He \rightarrow {}^{16}O$, ^{20}Ne, and ^{24}Mg, where n = 1, 2, or 3). These reactions are extremely temperature sensitive, scaling as T to the *fortieth* power (T^{40}) so that a 5 percent increase in temperature yields a 700 percent increase in burning rate and thus energy production. For stars with initial masses less than approximately eight times the mass of the Sun, the ultimate result is the same, however: pulsational instability, the ejection of the outer layers, and death, leaving a white dwarf corpse. Again, some of this newly synthesized material can enter space during the stars' death throes (because a white dwarf has to be less than 1.4 solar masses, the bulk of the stars' initial mass is lost in the process), but most of these elements are locked away, never again to participate in the cycle of stellar birth and death.

For stars beginning life with more than eight solar masses of material, however, life is much more exciting. As noted earlier in this chapter, despite having much more nuclear fuel than lower-mass stars, these massive stars use it profligately, generating energy at thousands to hundreds of thousands of times the rate of solar-type stars. Their lives are thus proportionally shorter—a 25-solar-mass star lives less than 7 *million* years.

Like all stars, the first step for even this 25-solar-mass behemoth is to fuse Hydrogen into Helium; this process starts when the core temperature reaches about 14 million K and lasts for about 6 million years. Next, as with the Sun, it's He produces C, which starts when the temperature T = 100 million K but lasts a mere 700,000 years. Then its Carbon becomes Oxygen, Neon, and Magnesium as in 3- to 8-solar-mass stars starting at 500 million K, but it doesn't stop there. Neon can burn to Magnesium too at 1.2 billion degrees. But by this time, the photons being produced have such high energies they can split a Helium nucleus off the Neon nuclei and generate more Oxygen:

$$^{20}\text{Ne} + \gamma \rightarrow {}^{16}\text{O} + {}^{4}\text{He},$$

which is how Oxygen gets to be the third most abundant element in the universe.

And the process continues. Oxygen and Neon become the fuel at 1.5 billion K and make Sulfur, Silicon, and Argon; this cycle is completed in just 3 months. Then, in the final day or two of the star's life, Silicon becomes the fuel (at T = 3 billion K) and makes Iron (plus some Chromium, and Nickel). Because Iron is just element number 26 in the Periodic Table, one might imagine the process can just keep going, making heavier and heavier elements. However, one Iron isotope, ^{56}Fe, is the most stable possible arrangement of protons and neutrons. Up to this point, every new nucleus produced was more stable than its predecessor, and thus the reaction gave off energy. Flowing out from the core, this energy held the star up against gravity. But when we get to ^{56}Fe, there is no more energy to be had—adding a neutron or a proton or a helium nucleus makes the new nucleus *less* stable, and thus it absorbs rather than emits energy.

For the star, the result is catastrophic. After millions of years during which its sequence of nuclear reactions generated outflowing energy to counter gravity's attraction, energy is suddenly sucked out of the center of the star as it tries to burn Iron into Cobalt. In less than 1 second, the core of the star containing 1 to 2 solar masses of matter implodes, going from a body the size of the Earth to a sphere the size of Manhattan. The collapse releases more energy in that 1 second than the Sun will produce in its entire 11-billion-year lifetime. And a lot of energy released in a given location at a particular moment of time is usually called an explosion—in this case, we call it a supernova: the outer layers of the star are blasted into space at 30,000 km/sec (roughly one tenth the speed of light, or New York to Sydney in half a second).

These events are responsible for distributing all the elements, from Beryllium to Iron, cooked up during the star's brief lifetime, out into interstellar space. Some of these nuclei are radioactive, as is the Aluminum isotope ^{26}Al that shaped the asteroids in our solar system, which we discussed in chapter 15, and Titanium-44 (^{44}Ti) that the NuSTAR satellite has detected at the site of recent stellar explosions.[5] Direct measurements of these isotopes' decay products and the detection in the X-ray band of electronic transitions from the other stable elements created throughout the star's lifetime provide unequivocal evidence that the atoms heavier than Lithium are made inside stars and distributed by supernova explosions.

As we have seen, it takes an input of energy to make the sixty-eight elements heavier than iron, but the supernova event has lots of extra energy, and some of those (much rarer) heavier elements are produced during the explosion itself. As a result, the clouds of interstellar gas from which new generations of stars form are now enriched with the entire set of chemical elements out of which planets and moons, comets and asteroids, sequoias and students are made. In Carl Sagan's immortal words, we are all, indeed, star-stuff.

OTHER SITES OF ELEMENT PRODUCTION

The energy released in the implosion of a massive star's core is one trigger for a stellar explosion. There is a second explosive mechanism, however, that can also both produce copious new elements and distribute them to interstellar space. As noted earlier, white dwarfs have a maximum mass limit—the pressure of all those electrons assigned to their unique quantum levels, the electron degeneracy pressure, is sufficient to hold up a star with a mass of less than 1.4 solar masses. The added gravitational attraction of just a little bit more mass, however, pushes the star over the edge of stability. Most white dwarfs are comfortably under this limit, so they just quietly cool and fade away. But if a white dwarf has a close companion star that can add mass until the dwarf hits the mass limit, fireworks result.

The majority of stars, unlike the Sun, do have one or more companion stars, so mass exchange between them is far from unknown. In most cases, however, the donor star lays down a layer of new material on the surface of the white dwarf until it gets to be a kilometer or so thick, at which point it triggers a thermonuclear explosion on the surface that, in addition to producing some new elements, blasts off a layer at least equal to, if not greater than, the amount of material stolen from the companion star in the first place, leaving the white dwarf's mass largely unchanged (or slightly shrinking). In some cases, however, the mass actually grows; whether this is from mass donated by a normal-star companion or from the merger of two white dwarfs in orbit about each other we have yet to determine. But when the 1.4-solar-mass limit is breached, it triggers a wave of nuclear burning that sweeps through the entire star, cooking up lots of new elements and blowing the star to smithereens.

While massive star supernovae produce lots of Oxygen and its burning products such as Neon, Magnesium, Sulfur, and Silicon, these thermonuclear supernovae produce more elements near Iron in the Periodic Table as fusion proceeds all the way to the most stable nuclear form throughout the star. And while small amounts of elements heavier than iron are produced in both types of events, we have long been uncertain about the origin of most of the heavier elements in the Periodic Table. Elements with atomic numbers greater than 26 are much less abundant than their lower-mass cousins (there are about 1 million Iron atoms and 10 billion Hydrogens for every one to ten atoms of elements 44 to 94, for example), so the processes that produce them can be suitably rare. A recent discovery using gravitational wave detectors[6] provides strong evidence for at least one new heavy-element source.

In describing the death of our 25-solar-mass star above, I left unremarked the fate of the imploding core. That sudden, dramatic collapse pushes quickly past the electron degeneracy threshold that halts a white dwarf's demise, driving all the electrons to fuse with the protons to form neutrons. Thus, a neutron star is born. If a white dwarf is like a giant macroscopic atom, a neutron star is a giant macroscopic atomic nucleus. It has the same density of a regular atomic nucleus, about 1 billion tons per cubic centimeter (that's *all* the cars, trucks, and SUVs in the world compressed to the size of a sugar cube) and is 10 kilometers (rather than 10^{-14} meters) across. The neutrons, being fermions (those particles whose spin = ½ and that resist close companionship; see chapter 3), follow an analogous rule to electrons and resist occupying identical quantum states—they just do so at much higher densities where "neutron degeneracy pressure" stops the otherwise inevitable gravitational collapse.

Neutron stars are remarkable objects that produce many of the most extreme phenomena we have discovered in the universe over the last half century. Indeed, I have managed to build an entire career out of studying these exotic stars, and so I could easily go on for a couple of hundred pages about them. Fortunately for you, dear reader, I will resist the temptation. However, a recent breakthrough regarding the creation of the heavy elements involves neutron stars, so a brief divertissement is required.

Neutron stars were predicted less than two years after the discovery of the neutron itself by Walter Baade and Fritz Zwicky, who wrote with remarkable prescience in 1934: "With all reserve, we advance the view that a supernova represents the transition of an ordinary star into a neutron star."[7] Thirty-four years later, Cambridge University graduate student Jocelyn Bell discovered a bit of "scruff" on a strip chart recorder attached to her radio telescope that had been designed to assess the twinkling of extragalactic radio sources as their signals passed through the interplanetary medium. When she sped up the recorder, she saw that the "scruff" was a series of pulses separated by the remarkably constant period of 1.33733 seconds. After firmly establishing that this signal came from the sky and briefly considering the little green men (LGM) hypothesis, the Cambridge team discovered three additional pulsed signals with different periods in different parts of the sky and concluded they must be a natural astronomical phenomenon. Within a year, rapidly rotating, magnetized neutron stars with lighthouse-like beacons sweeping over the Earth once per rotation were established as the explanation; these remarkable new denizens of our galaxy were dubbed pulsars. The discovery of the most rapidly spinning such star at the center of the remnant of the supernova of 1054 AD confirmed Baade and Zwicky's hypothesis.

When I began graduate school, I was assigned an office with an older student, Russell Hulse, with whom I shared Joseph Taylor as a dissertation adviser. The discovery of pulsars was only five years old at the time, and many searches were underway to find more of these exotic stars. Hulse and Taylor had designed the most sensitive search ever. It was to be conducted at the Arecibo Observatory, the world's largest radio telescope at the time. In particular, they were on a quest to find a neutron star with a companion. As noted above, it is common for stars to have companions and, in such cases, one can measure the masses of the two stars with considerable precision. Within a year, they hit paydirt, finding PSR 1913+16, a pulsar with a spin period of 0.059 seconds orbiting an unseen companion (almost certainly another neutron star) once every 7.75 hours. This allowed a mass measurement of the neutron star: 1.441 solar masses, above the maximum limit for white dwarfs, as expected.

More important, however, was that finding two such dense objects in such a tight orbit allowed for a novel test of the theory of general relativity. Einstein's revolutionary view of gravity, published in 1916, envisioned the mutual attraction of objects with mass as arising from warps in the fabric of space-time that the masses themselves produce. Imagine a trampoline (as space-time) with a bowling ball placed at the center (a massive object). A marble rolled across the surface of the trampoline will naturally fall toward the bowling ball, not because of some magical attraction between the two objects but because it is forced to travel along the curved space-time the bowling ball has induced. Furthermore, as objects move, they will send ripples through the space-time, like a stone dropped in a pond. Einstein had recognized the existence of the ripples his theory implied, but when he calculated their amplitude, he concluded they were far too weak ever to be detected. What he didn't take into account, however, was the stunning technological progress our species would make over the next century that would, among many other things, allow us to detected city-sized stars thousands of light-years away in space.

When the amplitude of the gravitational waves expected to be generated by the binary neutron star was calculated, the result suggested that the waves would drain energy from the system so that its 7.75-hour orbit should shrink by about 1 second over the ensuing four years and by even more as time went on. Indeed, this remarkable prediction was precisely borne out by the data; today, nearly fifty years after its discovery, the orbital period of the PSR 1913+16 system is 66 seconds shorter, consistent to better than 0.3 percent with Einstein's prediction.

When these observations were made in the 1970s, a handful of visionary scientists were already plotting how to build a device to detect gravitational waves directly. Their forty-year effort led to the Laser Interferometer Gravitational-wave Observatory (LIGO). With two stations in Louisiana and Washington State, this remarkable device can detect a change in the length of one of its 4-kilometer arms induced by a passing gravitational wave that amounts to one one-thousandth the width of a proton.

In August 2017, just a century after Einstein published his new theory of gravity, LIGO detected a rippling of its arm-lengths, beginning with a frequency of forty times per second and building in amplitude and frequency over 30 seconds to nearly 500 oscillations per second, at which point the signal disappeared. What they had witnessed was the final 1,500 orbits of two neutron stars that ended in their merger into a black hole—the ultimate fate of the Hulse-Taylor binary pulsar. Two seconds later, the Fermi Observatory orbiting the Earth detected a half-second burst of gamma rays. Within hours, telescopes around the world focused on a bright new "star" in a galaxy 130 million light-years away that marked the site of this cataclysmic event. Over the following month, the new star gradually faded from view, but not before we had recorded data that implied the creation of massive quantities of heavy elements—roughly 200 Earth masses of gold and 500 Earth mases of platinum, for example.[8] LIGO is currently undergoing an upgrade, and gravitational wave detectors in Europe, Japan, and India are coming online soon. Within a few years, we expect to observe one of these events every week or so, allowing us to measure quantitatively the major role that neutron star mergers play in the creation of the heavy elements.

There are other minor processes that produce elements. Material in disks swirling around supermassive black holes can generate some new nuclei, although, with at most one supermassive black hole per galaxy, this is not thought to be a significant source of new elements. Three of the lightest elements—Lithium, Beryllium, and Boron (numbers 3, 4, and 5, respectively)—are mostly produced by the shattering of heavier atomic nuclei in collisions with cosmic rays, super-high-energy particles that permeate intergalactic space, where they are accelerated in the remnants of exploded stars. With the exception of the primordial Hydrogen and Helium, the vast majority of our atomic historians are born in the death of stars—normal stars, white dwarfs, or neutron stars. Our final task, then, is to find out where the H and He come from. On to the Big Bang.

CHAPTER 17

In the Beginning

When we look out into space with a radio receiver tuned to wavelengths between 1 cm and 1 mm, we see energy coming at us from all directions. The radiation has a characteristic temperature of 2.725 K, just a bit above absolute zero. This celestial glow is remarkably uniform. Imagine a topographic map of the sky where the most intense parts are mountains and the faintest spots are valleys. We'll set the mean level as the height of the Empire State Building (1,450 feet from the sidewalk to the top of the antenna). In this cosmic map, the highest mountain would be an ant standing on top of the antenna and the lowest approximately 1 cm below the tip of the antenna. This signal, with its stunning uniformity overlain by such tiny fluctuations, represents a baby picture of the universe (figure 17.1).

This cosmic microwave background (CMB), discovered in 1965,[1] was first predicted twenty years earlier as the expected remnant of a hot primordial state of the universe we now call the Big Bang. It provides us with a wealth of information about what the universe was like in its infancy 13.8 billion years ago. Emitted just 390,000 years[2] after the moment our space-time came into existence, it also allows us to infer what conditions must have been like back to the very beginning of time itself.

The universe began very hot and very dense and has been expanding and cooling ever since. As the CMB photons move through this stretching space, the peaks of their waves get pulled farther and farther apart, increasing their wavelengths and lowering their energies. Today, their wavelength is about 1 millimeter; when they were emitted, it was more like 1 micrometer, 1,000 times shorter. Today, these waves of light are characteristic of a temperature of just under

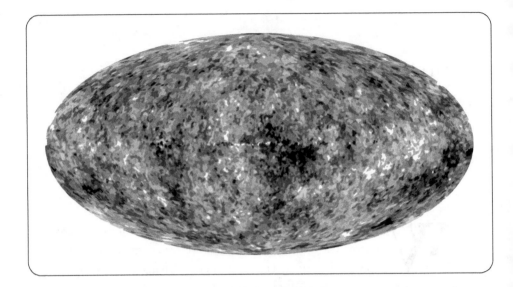

17.1 The spatial distribution over the entire sky of the cosmic microwave background radiation temperature. The darkest regions differ from the brightest regions by only about 4 parts in 100,000, but they represent the fluctuations in density and temperature that gave rise to the highly structured universe we see today.

3 Kelvins; when they started streaming toward us, the temperature of the universe was more like 3,000 Kelvins, 1,000 times greater.

Why, then, at that particular moment in the history of the universe? The photons were not created at that time, they already existed and earlier had even higher energies. But 3,000 K is about the temperature at which electrons can combine with protons and Helium nuclei to make Hydrogen and Helium atoms. When it is hotter than this, the ambient photons are energetic enough to ionize the electrons quickly, as soon as they join the nuclei, so stable atoms can't form. When the electrons are set free from their nuclei, they move very fast (because they are almost 2,000 times less massive than the nuclei) and are constantly running into photons and deflecting their paths. Thus, the transition from free electrons to those bound to nuclei suddenly makes the universe transparent, and the CMB photons can stream directly to us, where we catch them after their 13.8-billion-year journey to reveal the distribution of matter 390,000 years after the Big Bang.

There are lots of photons in the CMB. Every sugar cube–sized piece of space is occupied by 411 of them (yes, including in your backyard right now). They are all traveling at the speed of light (they are part of the electromagnetic spectrum after all), so when you walk out your door, about 50 trillion of them land on your head each second. You don't feel a warm glow because they are *very* low-energy photons; it would take 3 years of continuous exposure to equal the energy you receive in a second from a 1-watt Christmas tree light bulb. This vast number of photons, however, provides one important clue to the origin of everything. If we compare it to the number of protons (or electrons—they are equal in number) in the universe today, the answer is that there are 1.6 billion photons for each proton.

Unfortunately, we don't have a good way to predict this number, but it is straightforward to measure it. What it tells us is that when the universe began, it was almost perfectly balanced between matter and antimatter particles. For a reason we do not understand, the match wasn't quite perfect, however. In fact, there were 1,600,000,000 antimatter particles and 1,600,000,001 matter particles. The 1.6 billion of each met up and annihilated, producing all those photons; all the stars, planets, and galaxies in the universe today are made of the one-in-a-billion particles left over.

We cannot "see" directly what happened before the first atoms formed, but as Blaise Pascal said, "If our view be arrested there, let our imaginations pass beyond."[3] We can run the clock backward from 390,000 years ago when the temperature was about 3,000 K and there were about 1,000 atoms per cubic centimeter (about the density of a typical interstellar cloud today but far better than the best vacuum we can create on Earth). When we're 99.99999999 percent of the way back to the beginning, the universe is just three minutes old (yes, three of our minutes, 180 seconds).

The temperature is now 1 billion degrees (like that in the core of a massive star), but the density is only about ten times that of air. Between this moment and 1 second, our nuclear building blocks are all made. Protons, emerging from even earlier times, collide violently enough to stick together sometimes, making Deuterium (^2H), Helium (^3He and ^4He), and a tiny amount of Lithium (^7Li, and perhaps even a trace of ^6Li). Observations suggest about 24.5 percent of the matter becomes ^4He, .0035 percent is ^2H, .001 percent is ^3He, and 5 out of every 10 billion particles are ^7Li. These observations in turn allow us to infer the conditions when the universe was 1 second old—the matter density was about one-tenth that of water and the temperature was 10 billion degrees, the point at which photons lose the ability to form electron-positron pairs so that the ratio of photons to matter particles is fixed forever.

There is no need to stop here. At 10^{-4} seconds, the entire universe is only the size of our solar system, into which all of today's 10^{78} particles are packed. The temperature is 1 trillion Kelvin and the density is now that of an atomic nucleus. As we have seen, when quarks are packed this tightly together, the strong force kicks in and they fuse into triplets to make protons and neutrons. Before this moment, the photons are so energetic they can break these triplets apart, but as the universe "cools" to 1 trillion degrees, that is no longer possible—the protons and neutrons are secure and free quarks are banished from the cosmos.

Before even this, of course, stuff is still happening. At an age of 1 microsecond (10^{-6} s), the temperature is 10 trillion K and the density is 1,000 times that of an atomic nucleus. Before this, there are no composite particles—just the most basic building blocks of quarks, electrons, their antiparticles, and photons (plus a whole lot of fairly useless neutrinos). While cosmologists' models allow us to probe still further back, a universe the size of a large star—very hot, very dense, and consisting solely of structureless particles—will do for our purposes. It is likely here that the mysterious asymmetry between matter and antimatter unfolds, giving us the stuff out of which everything is made.

This model describes conditions very far from our quotidian existence, but it is not rampant speculation. It relies on the physics we know very well that has been tested thoroughly in our laboratories and particle accelerators. The fusion of quarks into protons and neutrons, the annihilation of matter and antimatter, the construction of nuclei beyond Hydrogen, and the formation of neutral atoms are all thoroughly understood phenomena. The temperature and tiny fluctuations of the CMB, the relative abundances of each of the light nuclei, and the large-scale structure of the universe today all provide tight constraints on the conditions that prevailed in the first moments of the cosmos. In particular, we understand the history of the quarks and leptons from $t = 10^{-6}$ seconds to the present, 13.8 billion years later. Equipped with that history, we have been privileged to use the atoms created from those elementary particles as historians to enlighten us about our world.

Epilogue

A Quark's Tale

Throughout this book, we have used the presence or absence or excitation or transformation of our tiny atomic historians to reconstruct the stories of our culture, our planet, and our universe. Yet each of these individual historians has a history of its own and, given the great stability of their constituents, that history extends back to the beginning of time. The following parable summarizes all that we have learned.

I am an UP quark. Emerging from the chaotic quantum soup of the universe when it is but one microsecond old, I spot a strongly attractive UP-DOWN pair. The three of us hit it off immediately, and we snap together to form an unbreakable bond, becoming one of the universe's first protons. One hundred long microseconds later, my proton collides head on with a somewhat-too-energetic electron, which gets absorbed, canceling my positive charge and forming a neutron. At $t = 30$ seconds, bored in this neutral state, my companion quarks and I spit out the electron again (i.e., our neutron undergoes beta decay), and we return to the essence of proton-ness.

Within a few seconds, we collide with another neutron that sticks, forming a deuterium nucleus. Alas, just seconds later, a photon smacks into us (it's a very crowded party) and splits us apart again. Within a minute, however, we find another neutron and this time the relationship lasts because, given the ongoing expansion of the universe and the accompanying stretching of all the photons' wavelengths, none of them now has the energy to rend us asunder. We make such an attractive couple that we become a party magnet; within seconds another deuterium nucleus arrives, and we quickly form a nuclear family of four—a stable Helium nucleus.

Hundreds of thousands of years pass, but nestled in our helium nucleus, I remain positive. The party has thinned out drastically, and it feels distinctly cooler. After 390,000 years, it's down to a few thousand degrees out there, and when a couple of electrons wander by, we captivate them, and they promise to orbit us forever. One of them sneaks away occasionally after getting too close with a particularly energetic photon, but because all electrons are identical, another one soon relaces it and no one is the wiser. Our Helium atom is content with its two electrons and is unwilling to form bonds with other atoms (really, it's not unwilling, but unable; however, we prefer to adopt an attitude involving self-determination). We drift around alone in the ever-expanding universe.

After about 400 million years, we find ourselves in a cloud of fellow Hydrogen and Helium atoms that is inordinately attractive and is in the process of recruiting about 10^{68} fellow atoms to form this huge structure that will one day be called the Milky Way. Our nearest atomic neighbors buzz around just a centimeter or so away. After another 9 billion years has passed, the neighborhood gets ever more crowded, first with two, then ten, then ten thousand, then 1 million companions crammed into the sugar cube–sized bit of space we used to have all to ourselves!

And then it gets much worse. We find ourselves falling and compressing and falling and compressing until the collisions with our neighbors are so violent that our electrons are stripped away, gone forever. Soon we are colliding with other nuclei millions of times each second. All around us, protons are hooking up to make Deuterium, and Deuteriums are connecting to make our wimpy cousins Helium-3's, which in turn combine to make more regular Helium nuclei like us. Eventually, that's all we see around us—trillions of identical cousins. Suddenly, it gets hotter and denser still, and we collide so violently with two of our brethren that we can't get away—we are bound together in a Carbon nucleus.

A few hundred thousand years later, we are suddenly thrust out of this seething cauldron, shot out at a tenth the speed of light into the cool darkness of interstellar space. It is now just too cold—only about 10 degrees above absolute zero, and the slowly moving neighbors start crowding together again. After 1 million years, we collide with a tiny grain of dust and stick to its surface. Now we're on solid ground, and it is time to check out the neighbors. Collecting five other Carbons, two Nitrogens, two Oxygens, and fourteen Hydrogens for drinks allows us to form long-lasting bonds (in a left-handed sort of way), making lysine.

About 10 million years elapse, and our little grain lands with a thud on a two-kilometer-wide asteroid. We see lots of other rocks and boulders flying by, and after another 20 million years, we fall hard onto the crust of the newly formed place called (for some unknown reason) Earth. Things finally start to calm down—seems like I may have found a permanent home at last as an UP quark in a proton in a Carbon nucleus in a lysine molecule on a cooling planet.

Then, one day, I find my home molecule is floating in a pool of water. A few adenine molecules swim by and when they see us, one of them grabs on tight. Then, as if by some unseen plan, more and more amino acids like us join a chain, and the chain starts folding into all kinds of contorted shapes until we find ourselves stuck into the cell wall of an Archaea. But there is no rest for the weary. The Archean cell soon gives up the ghost and sinks to the bottom of a pond where it gets ingested by a worm, which gets eaten by a bird, and I end up in an egg yolk—and so it goes for a billion years.

At some point, my Carbon atom host breaks free from its linked companions, and we're off to find new friends in the calcium carbonate ($CaCO_3$) shell of a foraminifera floating in the ocean—floating, that is, until it dies and sinks to the floor of the ocean four miles below. Things move very slowly down there, but move they do and eventually I feel it getting hot under our feet and we are suddenly and unceremoniously shot out of a volcano, linked to two Oxygen friends and free to roam around in the atmosphere. Once, I got a little too close to the ground in winter and got trapped in a bubble of air encased in ice for 300,000 years before being liberated into a cold lab by a curious woman with a big needle.

We escaped out the window but were soon captured again by a tree leaf, where my Oxygen friends were rather rudely ripped off, and I was able to join some identical Carbon siblings in a chain of cellulose. This was peaceful for 100 years or so, until we were rousted out by a man with a drill and dropped, unnoticed, in a speck of wood on the forest floor. A passing turkey, pecking at the ground for seeds, accidentally ingested us, and I soon found myself again in one of those long, folding chains in the turkey's thigh. That lasted only about six months, until it got quite hot for a few hours and then, to the accompaniment of chewing sounds and loud arguments, we got rearranged once more, reconnected with two Oxygen friends, and set free by a gentle breath, back into the atmosphere to continue the journey. We UP quarks get around.

My quark has been on its journey for 13.8 billion years. The system we have for uncovering my story—science—has only been around for a little over

400 years, since the January evening when Galileo Galilei turned his newly made telescope to the night sky and began the process of testing conflicting hypotheses, not through the elegance of the rhetoric with which they were expressed but by observation and experimentation. Galileo knew nothing about atoms, but he gave us a new approach to the universe—that it is a great book "written in the language of mathematics" and posited that we can come to understand this book if we learn its language: in short, that the universe—the UNIty of the diVERSE— is comprehensible.[1]

It is with Galileo's spirit that we have applied our understanding of the physics of atoms to matters art historical and archeological, chemical, geological, and astronomical in order to enrich and extend our view of history. A few of our topics have practical applications—verifying art works, diagnosing medical conditions, and probing the extremes of past climates to help predict our future. Others, such as knowing when indigenous peoples learned to plant corn or how exactly the Carbon atom in your toenail was cooked up inside a star, will affect neither your longevity nor your brokerage account. Nonetheless, it is my hope that these tales of atomic detective work have enhanced your appreciation both for history and for science by expanding your perspective on our place in the universe.

Glossary

While I have attempted to define carefully all of the scientific terms used in the text, it would be unsurprising if, by chapter 13, the reader had forgotten exactly what an ion is or how isotopic fractionation works. Thus, I provide here, for easy reference, a glossary of terms with what I hope are memory-jogging, brief definitions; in most cases, greater elaboration can be found in the text.

ACCUMULATION CLOCK A (somewhat simplistic) radioactive dating technique that assumes a sample initially contains none of the decay sequence's daughter isotope and that none of the daughters created in the decay leak out of the sample. In this case, the ratio of parent to daughter isotopes gives the sample age directly (see figure 6.5).

ALPHA-DECAY The radioactive transformation of an atomic nucleus involving the emission of a Helium nucleus (two protons plus two neutrons). This type of transformation changes the atomic number of a nucleus by −2 and the atomic mass by −4.

ANTIMATTER The mirror versions of elementary particles that are identical to normal particles except that their electric charges are reversed. The anti-electron (or positron) has a charge of +1, while the anti-up quark has a charge of −⅔. The antineutron retains its charge of zero, but instead of udd, it is anti-u, anti-d, anti-d. Antimatter particles are represented by their particle symbol with a horizontal line on top (e.g., an antineutron is $\overline{udd}$).

ATOMIC MASS The sum of the number of protons and neutrons in an atom. It is not equal to a mass measured in kilograms because some mass-energy is lost in the fusion of the nuclear particles that make the atom.

ATOMIC NUMBER The number of protons (and, thus, for a neutral atom, also the number of electrons) in an atom. All atomic numbers from 1 to 118 are represented in the Periodic Table of the Elements.

ATOMS The basic building blocks of all matter consisting of a massive, positively charged nucleus made of protons and neutrons surrounded by a cloud of light, negatively charged electrons.

BETA DECAY The radioactive transformation of an atomic nucleus involving the emission or absorption of an electron or its antiparticle, the positron. This type of transformation changes the atomic number of a nucleus by ±1 but leaves the atomic mass unchanged.

BOSON The class of elementary particles with quantum spin 0, 1, or 2. Photons, gluons, and other force-carrying particles are bosons, as are any composite particle such as Helium-4 and other stable nuclei with an even mass number.

CHARGE A fundamental property of matter that gives rise, and responds, to the electromagnetic force.

CHEMICAL REACTION A transformation involving the outermost electrons of atoms in which they combine or break apart from their relationships with other atoms, either absorbing or releasing energy. Typical energies are within a factor of 10 of 1 eV per atom.

CONCORDIA A radioactive dating technique that takes into account the possible escape of daughter nuclei from the sample under study (see figure 12.1).

ELECTROMAGNETIC SPECTRUM The range of wavelengths of self-reinforcing waves of electric and magnetic energy that move through empty space at the speed of light (3×10^8 m/s). Visible light represents one octave of the spectrum, which ranges over 60 octaves in nature, from radio waves through microwaves, infrared rays, ultraviolet rays, X-rays, and gamma rays.

ELECTROMAGNETISM One of the four forces of nature that causes objects with the property of charge to attract (for opposite charges) or repel (for like charges) each other. The photon is the carrier of electromagnetism.

ELECTRON An elementary particle belonging to the fermion class and bearing an electric charge of −1.

ELECTRON-VOLT (EV) A unit of energy appropriate for interactions of fundamental particles on the atomic level, where $1 \text{ eV} = 1.6 \times 10^{-19}$ Joules. Interactions between atoms in molecules typically have energies of 0.1 to 10 eV, electrons are held to nuclei by energies between a few and a few hundred thousand eV (100 keV), and nuclear processes typically have energies of millions of eV (MeV).

ELEMENTS The 118 (1 to 94 natural; 95 to 118 humanmade) basic building blocks of all substances. They are defined by their atomic number (the number of protons; all versions of 1 to 118 exist).

ENERGY A construct representing the ability to do work—to change the amount or direction of motion. Its usefulness derives from the fact that, while there are many forms energy can take, it is never created or destroyed. The basic metric units are the calorie, which represents the amount of energy needed to raise 1 gram of water 1 degree Celsius, and the Joule; 4.184 Joules = 1 calorie.

FERMION The class of elementary particles with quantum spin of ±½. Electrons, neutrinos, quarks, and their respective antiparticles are all fermions, as are odd-numbered nuclei.

FORCE A push or pull. There are four fundamental forces of nature: gravity, electromagnetism, weak nuclear, and strong nuclear.

FISSION The splitting of an atomic nucleus either spontaneously (rarely) or through the collision with an external neutron or gamma ray. For the heavy elements, this process creates two lighter element nuclei and releases energy.

FRACTIONATION The process in which a chemical reaction either discriminates against or favors one isotope of an element over another, leading to an isotopic signature in the product differing from that in the pool from which the atoms were drawn. A classic example occurs in photosynthesis, in which plants that employ different photosynthetic pathways discriminate differently against the heavier Carbon-13 isotope compared with Carbon-12. Different kinds of plants are thus said to have different fractionation factors.

FUSION The merging of two atomic nuclei to create a heavier nucleus. For elements lighter than Iron-56, this process releases energy, typically tens to hundreds of MeV. This is the process that powers the stars.

GAMMA DECAY The release of gamma ray photons from an energetically excited nucleus. This often follows an alpha or beta decay that has transformed the nucleus and left it in an excited state. Gamma decay does not change either the atomic mass or atomic number of a nucleus.

GAMMA RAY A high energy photon ($E > 10^5$ eV).

GAS A collection of particles of the same or different kinds (the latter is called a mixture) in which the particles are free to move independently of one another, colliding elastically (like billiard balls) with each other and with the walls of their container. In the air at Earth's surface, the particles are separated by about ten times their individual diameters. Steam is the gaseous state of H_2O.

GLUON The force-carrying particle (a boson) of the strong nuclear force that keeps quarks bound into protons and neutrons and keeps atomic nuclei from flying apart.

GRAVITY The weakest of the four forces of nature (10^{-36} times that of the electromagnetic force) that causes all objects with the property called mass to attract each other. In Albert Einstein's picture, the presence of mass warps the fabric of space-time, which in turn constrains how objects move.

HALF-LIFE The time it takes for half a sample of radioactive atoms to decay (or, equivalently, the time span over which the probability for decay of a single radioactive nucleus is 50 percent). Half-lives range from tiny fractions of a microsecond to greater than the age of the universe. The half-life of a given radioactive nucleus is impervious to change by external forces, rendering it an imperturbable clock.

ION An atom that is electrically positive or negative as a consequence of having lost one or more electrons or having gained extra electrons.

ISOTOPE The various versions of a given type of atom (element) that differ by the number of neutrons they have. Such atoms are all chemically equivalent, but they have different masses and thus can be incorporated into chemical reactions at different rates.

ISOTOPE RATIO $^{13}C/^{12}C$—the number of C-13 atoms compared to the number of C-12 atoms. This is often quoted compared to a standard, such as the ratio in some substance compared to the ratio in the air.

JOULE A metric unit of energy equal to 1 kg m^2/s^2. One watt is the expenditure of 1 joule/second.

LEPTON The set of "light" elementary particles that includes the electron (and its heavier cousins the muon and the tau) along with their respective neutrinos and antiparticles.

LIQUID A collection of particles of the same or different kinds (the latter is called a mixture) in which the particles are touching each other but are free to slide around, such that a liquid takes the shape of the container into which it is poured. Water is the liquid state of H_2O.

MOLECULES Particles containing a fixed number of one or more elements bound together by relatively weak electrical forces exchanged between their outermost electrons. The Oxygen we breathe in consists of two atoms of Oxygen stuck together in an Oxygen molecule (O_2). The carbon-dioxide we breathe out is one Carbon atom joined to two Oxygen atoms (CO_2). Molecules can range in size from 2 atoms to greater than 10 billion atoms.

NEUTRINO An extremely low-mass particle (< 1eV or < 1/500,000 the mass of an electron) that is involved in reactions that in turn involve the weak nuclear force.

NEUTRON A neutral particle found in the atomic nucleus with a mass slightly greater than the proton and composed of a triplet of quarks (udd).

NUCLEUS The positively charged, massive core of the atom with a characteristic size of approximately 10^{-15} m, comprised of protons and neutrons.

PERIODIC TABLE The summary compilation of the 118 known elements. The rows correspond to the primary electron energy levels (n), and the columns group together atoms with similar external electron configurations (and thus similar chemical properties).

PHOTON A small packet of electromagnetic radiation with a mass of precisely zero and an energy inversely proportional to its wavelength.

POSITRON The antiparticle of the electron with a charge of +1.

PROTON A positively charged particle found in the atomic nucleus with a mass slightly less than the neutron and composed of a triplet of quarks (uud). The number of protons uniquely specifies an element's identity.

QUANTUM NUMBERS Numbers corresponding to an elementary particle's energy (n), angular momentum (l), and spin (s). In an atom, no two particles can have the same set of quantum numbers; this exclusion property of elementary particles is responsible for the electron energy configuration of atoms with their shells and subshells.

QUARK A group of elementary particles (fermions) that, in the universe today, are always bound up in pairs (mesons), triplets (protons and neutrons and a host of unstable configurations), or higher-order groupings. The Up (charge $+\frac{2}{3}$) and Down (charge $-\frac{1}{3}$) quarks make up protons (uud) and neutrons (udd).

RADIOACTIVE ISOTOPE An isotope that undergoes decay at a fixed rate; thus, it can be used like a clock. ^{14}C is radioactive; half of its atoms decay in 5,730 years.

RADIOACTIVITY The spontaneous and imperturbable transformation of an atomic nucleus into a different (or at least a less excited) nucleus.

SOLID A collection of particles of the same or different kinds (the latter is called a mixture) in which the particles are touching each other and locked in place so they are not free to move. Ice is the solid state of H_2O.

STABLE ISOTOPE An isotope that does not undergo radioactive decay, at least on a timescale shorter than one trillion times the age of the universe. It has a copacetic balance of neutrons and protons in its nucleus. There are 254 naturally occurring stable isotopes.

STRONG NUCLEAR FORCE One of the four forces of nature that acts only on quarks and their property of color charge, and only on scales $< 10^{-14}$ meters (i.e., inside the atomic nucleus). The strong force is about 100 times stronger than the electromagnetic force and is responsible for holding the atomic nucleus together.

TEMPERATURE A measure of the average kinetic energy of a collection of particles. The zero point on the absolute temperature scale of Kelvins, when all motion ceases, is at $-273.15°C$.

WEAK NUCLEAR FORCE One of the four forces of nature that acts only on scales $< 10^{-14}$ m (i.e., inside the atomic nucleus). It largely mediates radioactive decay and is about 1/10,000th the strength of the electromagnetic force, or 10^{-6} of the strong force.

X-RAY A relatively high-energy electromagnetic photon with an energy between roughly 100 eV and 100,000 eV.

Notes

INTRODUCTION

1. Originally, the second was defined as 1/86,400 of a day (24 hours × 60 minutes/hour × 60 seconds/minute = 86,400 seconds). However, because the length of the day depends on the rotation rate of the Earth and that rate changes with time, a new definition based on a hyperfine transition of Cs-133 was adopted in 1967. Current atomic clocks are accurate to 1 second in 3 billion years, roughly one quarter the age of the universe.

2. CONCEPTUALIZING THE ATOM: FROM PHILOSOPHY TO SCIENCE

1. Bhaskar Jha, "A Critical Study About the Nyaya-Vaisesika Theory of Atomism," *International Journal of Research and Critical Reviews* 5, no. 3 (2018): 920–923.

2. This is an open access, fee-to-publish journal; it does not appear on the list of suspect predatory journals kept at https://predatoryjournals.com/journals/, so I am assuming it is a legitimate source.

3. Falsifiability was introduced as a criterion for legitimate scientific work by the philosopher Karl Popper in *Logik der Forschung* (1934).

4. Stephen Greenblatt's *The Swerve: How the World Became Modern* (New York: Norton, 2012) won both the Pulitzer Prize and the National Book Award. The quote appears on p. 187.

5. The remarkable story of this loss and resurrection is brilliantly told in Greenblatt's *The Swerve*, cited in note 4 above, although it must be noted that while Greenblatt's discovery story has not been challenged, his characterization of the Middle Ages has been roundly criticized (e.g., J. Hinch, "Why Stephen Greenblatt Is Wrong and Why It Matters," *Los Angeles Review of Books*, December 1, 2012, and references therein).

6. Greenblatt, *The Swerve*, 220.

7. Saul Fisher, "Pierre Gassendi," in *The Stanford Encyclopedia of Philosophy*, ed. Edward N. Zalta, Stanford University, 2014, sec. 10, https://plato.stanford.edu/archives/spr2014/entries/gassendi/.

8. G. Schilling, *Ripples in Spacetime: Einstein, Gravitational Waves, and the Future of Astronomy* (Cambridge, MA: Harvard University Press, 2017).

9.   Antoine Lavoisier, *Traité élémentaire de chimie, présenté dans un ordre nouveau et d'après les découvertes modernes* (Paris: Chez Cuchet; Bruxelles: Cultures et Civilisations, 1965). The English-language translation of the title is "Elementary Treatise on Chemistry, Presented in a New Order and Alongside Modern Discoveries."

10. M. Bachtold, "Saving Mach's View on Atoms," *Journal for General Philosophy of Science/ Zeitschrift für allgemeine Wissenschaftstheorie* 41, no. 1 (June 2010): 1–19.

11. B. J. Ford, "Brownian Motion in Clarkia Pollen: A Reprise of the First Observations," *The Microscope* 40, no.4 (1992): 235–241.

12. S. Greenblatt, *The Swerve*, xx. It should be noted that the dancing of dust motes in a sunbeam is not exactly the same as Brown's microscopic motion of pollen grains because the latter motion is indeed caused by the statistically random collisions of individual molecules, whereas the macroscopic dust particles are in motion because of air currents that represent the coordinated directional motion of trillions of air molecules; qualitatively, however, the analogy is reasonable.

13. J. J. Thompson, "Cathode Rays," *The Electrician* 39 (1897): 104. This paper provided the first experimental evidence that atoms were indeed not "uncuttable" but were made up of smaller entities—Thompson called them "corpuscles"—with negative charge (they are now called electrons). He later envisioned the atoms as a "plum pudding" in which the electrons were like raisins in an amorphous distribution of balancing, positively charged dough.

14. Ernest Rutherford, "The Scattering of α and β Particles by Matter and the Structure of the Atom," *Philosophical Magazine* 21, no. 125 (1911): 669–688. This is the paper in which Rutherford reports the astonishing result of the experiment he carried out with Ernest Marsden, in which they fired alpha particles (now known to be Helium nuclei) at thin gold foils and found that some of them reversed direction and came back at the source. Rutherford famously noted, "It was as if you fired a 15-inch shell at a piece of tissue paper and it came back and hit you." The only plausible explanation was that the positive charge of the alpha particles was repulsed by a positive charge in the atom that, unlike the hypothesized, spread-out distribution of Thompson's atomic model, was concentrated in a tiny fraction of the atom's size—what we now call the nucleus.

15. Niels Bohr, "On the Constitution of Atoms and Molecules, Part I," *Philosophical Magazine* 26, no. 151 (1913): 1–24. In this paper, Bohr lays out his model of the atom that takes into account Rutherford's compact nucleus and introduces the notion of the quantum into atomic structure.

16. Max Planck, "On the Theory of the Energy Distribution Law of the Normal Spectrum," *Verhandlungen der Deutschen Physikalischen Gesellschaft* 2 (1900), 237 [English translation from *The Old Quantum Theory,"* ed. D. ter Haar (Pergamon Press, 1967), 82]. In this paper, at the dawn of the new century, Planck introduces the notion of a "quantum" of light, which we now call the photon.

17. Albert Einstein, *"Über einen die Erzeugung und Verwandlung des Lichtes betreffenden heuristischen Gesichtspunkt"* ["On a Heuristic Point of View about the Creation and Conversion of Light"], *Annalen der Physik* 17, no. 6 (1905): 132–148. This paper, explaining how light liberates electrons from atoms, is the first of the four papers in Einstein's *annus mirabilis*, 1905, and the one for which he was awarded the Nobel Prize in 1921.

3. THE ATOM: A UTILITARIAN VIEW

1. There is a fourth state of matter that in fact makes up most of the universe but is very rare on Earth: plasma. Plasma is found only in regions where the temperature is greater than about 10,000°C. Plasma consists of matter in which the electrons have been stripped from their parent atoms.

2. If you don't believe me, take a jar lid the size of your coffee cup and try to compress the liquid into a smaller space—but wait until the coffee has cooled to avoid first-degree burns.

3. The particles of air in your room are moving remarkably rapidly. The relationship between temperature and velocity is $v_{(peak)} = 145$ m/s sqrt $\{T[K]/m[amu]\}$ where $v_{(peak)}$ is the velocity of particles at the peak of the distribution (i.e., the most common velocity), T is the temperature (in Kelvins—explanation in the text, and m is the mass of the particle in atomic mass units (see chapter 4). Thus, the air particles in a room at 68°F are moving at nearly half a kilometer per second (1,100 miles an hour).

4. The temperature at which water boils depends mostly on the speed of its particles (i.e., its temperature) but also on the external pressure it is feeling from the air. It takes less energy for water molecules to escape from a liquid when there are fewer air particles knocking them back into the liquid state. Thus, the boiling point of water falls by about 0.5°C (0.9°F) for every 150 meters (approximately 500 feet) of altitude; in Denver, the Mile High City, water boils at 203°F (95°C).

5. If we assume that the bath water is at a higher temperature than the ambient air, every time an air particle collides with a water particle, it gets a kick and carries away some extra energy. This means the air is slightly warmer and the bath water is slightly cooler. This continues until the two reach an equilibrium in which each air-water collision is just as likely to transfer energy to the water as from the water—the water has become "room temperature."

6. The atmosphere is made of particles with mass, and everything with mass feels the force of gravity; thus, the atmosphere *does* in fact fall down. However, as described in note 3 above, the air particles are moving fast—on average, approximately 500 meters/second, with the fastest ones moving five or ten times that rate. Thus, they can reach high altitudes before slowing to a stop and falling back toward Earth. This explains why it is harder to breathe at high altitudes—only the fastest particles make it up that high and so there are fewer particles of Oxygen to breathe.

7. The bicycle pump gets hot because you are speeding up the air particles in it and, as I keep saying, higher speeds equal higher temperatures.

8. The air conditioner cools the room by bringing the rapidly moving (hot) air particles into contact with slow-moving coolant particles. The air particles lose some of their energy in the interaction, thus cooling the room.

9. You probably learned in school that there are ninety-two naturally occurring elements. I use ninety-four throughout this book because both neptunium (number 93) and plutonium (number 94) are found in trace amounts in rocks. In reality, however, this number is somewhat arbitrary. There are only eighty elements (see chapter 4) that have at least

one stable isotope (see chapter 5); the other twelve or fourteen (or even sixteen because numbers 94 to 98 also occur fleetingly on Earth) are the products of radioactive decay (chapter 6). In addition, it is probable that all of these, plus even heavier elements, are produced in the violent events—supernova explosions and neutron star mergers (chapter 16)—in which all heavy elements are formed, but these all have such short lives that they are not found in the Earth's crust. Nonetheless, they could be dubbed "naturally occurring."

10.   Even this number of thirty-one is somewhat arbitrary. For example, as noted in the text, the quarks have an additional property of "color charge," of which there are three varieties and that, if counted separately, would take the twelve quarks to thirty-six. There is also an outstanding question regarding neutrinos: are they are their own antiparticles (which would reduce my count by three)? The main point remains, however, that the current standard model of particle physics has an embarrassingly large number of supposedly "fundamental" particles.

11.   String theory presents an elegant mathematical structure that envisions a ten- or eleven-dimensional space in which the entities we regard as particles in our four-dimensional space-time are manifestations of the different vibrational modes of tiny "strings." Because this mathematics has yet to make falsifiable predictions about the material world, it is arguable whether it should be called science; passionate advocates on both sides of this question are numerous.

12.   Nuclear sizes range from the single proton of a Hydrogen nucleus with a radius of 1.7×10^{-15} m, to Uranium with 238 protons and neutrons packed into a space less than seven times bigger: 11.7×10^{-15} m.

13.   The word "element" was originally used to represent one of the four polyhedra posited to represent earth, air, fire, and water, of which everything else was thought to be composed.

4. THE ELEMENTS: OUR COMPLETE SET OF BLOCKS

1.   From Plato's *Timaeus* (approximately 360 BC) in which he adopts Empedocles's model from the previous century.

2.   J. R. Partington, *A Short History of Chemistry* (New York: Dover, 1937). A Hydrogen-filled dirigible with a passenger cabin underneath attempted a landing at Lakehurst, New Jersey, after a transatlantic flight from Germany and burst into flames, killing thirteen of the passengers, twenty-two of the crew, and one of the landing team on the ground.

3.   These are the so-called noble gases, which cannot form bonds with other elements (or even among themselves, unlike human nobility) because their outermost electron shells are full.

4.   An octave represents the interval between two frequencies separated by a factor of 2; for example, the A above middle C represents a vibration of 440 oscillations per second, whereas the A below middle C is 220 cycles per second, one octave lower. A string's length, thickness, and tension determine the rate at which the string vibrates, but for a fixed tension and thickness, doubling the length produces a sound exactly one octave below the original note.

5. Calorie with a capital C (the version used in food packaging) is a kilocalorie, or 1,000 calories with a small c. A calorie is the energy required to raise one gram of water by 1°C.

5. ISOTOPES: ELEMENTAL FLAVORS

1. The term "isotope" was suggested to the chemist Frederick Soddy by a family friend, Margaret Todd, one of the first matriculants to the Edinburgh School of Medicine for Women (1886).

2. Element 83, Bismuth, has an isotope with 126 neutrons and was long thought to be the heaviest stable isotope. In 2003, it was discovered that it could undergo alpha decay (see chapter 6). The half-life of Bismuth-209 is 1.9×10^{19} years, or 19 quintillion years, about 1.5 billion times the age of the universe.

3. Radioactivity, described in chapter 6, is the process by which atomic nuclei spontaneously transform themselves from one element or isotope to another.

4. The eight elements discovered in Berzalius's laboratory were Cerium, Selenium, Zirconium, Lithium, Lanthanum, Vanadium, Silicon, and Thorium. The last two had been named by others, but Berzelius was the first to isolate them and measure their properties.

5. Brittanica.com, "The Discovery of Isotopes," accessed February 2, 2023, https://www.britannica.com/science/isotope/The-discovery-of-isotopes.

6. F. W. Aston, "A Positive Ray Spectrograph," *London, Edinburgh, and Dublin Philosophical Magazine and Journal of Science* 38 (1919): 707–714.

7. Note that $35 \times .7577 + 37 \times .2423 = 35.485$, not 35.45, but recall that mass is lost to the nuclear binding energy in the creation of nuclei. See chapter 3.

8. The remaining elements with a single stable isotope are Phosphorus (number 15), Scandium (number 21), Vanadium (number 23), Manganese (number 25), Cobalt (number 27), Arsenic (number 33), Rubidium (number 37), Yttrium (number 39), Niobium (number 41), Rhodium (number 45), Indium (number 49), Iodine (number 53), Cesium (number 55), Lanthanum (number 57), Praseodymium (number 59), Europium (number 63), Terbium (number 65), Holmium (number 67), Thulium (number 69), Lutetium (number 71), Rhenium (number 75), and Gold (number 79).

9. The mineral from which Uranium is mined, now called uraninite, was originally dubbed pitchblende. The element Uranium was first isolated from it in 1789. Marie Curie had to process tons of pitchblende to discover Radium in 1910. The tiny concentrations of isotopes such as Technetium, Promethium, and Francium found in uraninite are the products of the spontaneous fission of Uranium (see chapter 6); Lead and Helium are also found as products of the elements' radioactive alpha decays.

10. The exception is the CANDU reactor developed in Canada as a pressurized water reactor that uses Deuterium Oxide (heavy water in which the Hydrogen atoms are replaced with Deuterium or ^{2}H) as the moderator instead of the Boron or Carbon (graphite) used in most commercial nuclear power plants. The CANDU reactors use natural Uranium ore where the ^{235}U fraction is only 0.7 percent, thus obviating the need for enrichment (and the weapons potential inherent in that step).

11. H. L. Rosenthal, J. E. Gilster, and J. T. Bird, "Strontium-90 Content of Deciduous Human Incisors," *Science* 140, no. 3563 (1963): 176–177.

6. RADIOACTIVITY: THE IMPERTURBABLE CLOCK

1. Henri Becquerel, "Sur les radiations invisibles émises par les corps phosphorescents," *Comptes Rendus* 122 (1896): 501–503, translated by Carmen Giunta, accessed March 2, 2019.

2. I cannot resist noting here that while the public is deathly afraid of nuclear power-plant waste because of the small volumes of long-lived radioactive isotopes such plants produce, they have been blissfully unconcerned for more than two centuries about the far more damaging waste from burning fossil fuels—the carbon dioxide, methane, and nitrous oxides that are transforming Earth's atmosphere and disrupting our climate. In fact, while the most extreme estimates of deaths from nuclear power plants over the past 65 years number in the low thousands, the *annual* death toll for fossil fuel power-plant pollution exceeds 3 million, not even counting lives lost from climate disruption.

3. B. Wang et al., "Change of the ^{7}Be Electron Capture Half-Life in Metallic Environments," *European Physical Journal A—Hadrons and Nuclei* 28 (2006): 375–377.

7. STOLEN AND FORGED: FORENSIC ART HISTORY

1. F. Caro and I. M. Sokrighy, "Khmer Sandstone Quarries of Kulen Mountain and Koh Ker: A Petrographic and Geochemical Study," *Journal of Archeological Science* 39 (2012): 1455–1466.

2. The stories of the Spanish Forger and the Blakelock forgeries in this chapter are based on the experiments recounted in Maurice Cotter, "Neutron Activation Analysis of Paintings," *American Scientist* 69, no. 1 (1981): 17–27, and references therein.

3. After twenty half-lives, the number of radioactive nuclei remaining would be $(\frac{1}{2})^{20} = 9.5 \times 10^{-7}$, or just under one in a million.

4. This pigment is so toxic it is no longer used. During Napoleon's exile on St. Helena, he lived in a house where the walls were painted with his favorite color, bright green. Analysis of his hair postmortem revealed high levels of arsenic, which has been linked to stomach cancer, the proximate cause of his death.

5. Cotter, "Neutron Activation Analysis of Paintings."

6. Wikipedia, s. v. "Ralph Albert Blakelock," accessed February 2, 2023, https://en.wikipedia .org/wiki/Ralph_Albert_Blakelock.

7. H. Smith, "Genius in the Madhouse," in *The Saturday Review*, March 31, 1945, reprinted in *The Saturday Review 50th Anniversary Reader: The Golden Age*, ed. Richard L. Tobin and S. Spencer Grin (New York: Bantam Books, 1974).

8. H. S. Ching, Z. Q. Zhang, and F. J. Yang, "Non-Destructive Analysis and Identification of Jade by PIXE," *Nuclear Instruments and Methods in Physics Research Section B: Beam Interactions with Materials and Atoms* 219–220 (2004): 30–34.

9. P. Del Carmine, L. Giuntini, W. Hooper, F. Lucarelli, and P. A. Mando, "Further Results from PIXE Analysis of Inks in Galileo's Notes on Motion," *Nuclear Instruments and Methods in Physics Research Section B: Beam Interactions with Materials and Atoms* 113 (1996): 354–358.

8. THE CARBON CLOCK: PINNING DOWN DATES

1. The headline from B. Booker, "Carbon Dating Suggests Early Quran Is Older Than Muhammad," *Digital Journal*, August 31, 2015, https://www.digitaljournal.com/world/carbon-dating-suggests-early-quran-is-older-than-muhammad/article/442550, is one of many examples with very similar wording appearing in the world's media in the summer of 2015.

2. The current location and rate of change of the North Magnetic Pole's location, tracked by the National Oceanic and Atmospheric Administration, can be found here: https://www.ngdc.noaa.gov/geomag/GeomagneticPoles.shtml.

3. Data on the strength of the Earth's magnetic field from the European Space Agency can be found at https://www.esa.int/Applications/Observing_the_Earth/Swarm/Swarm_probes_weakening_of_Earth_s_magnetic_field.

4. A description of the so-called bomb pulse of C-14 in the atmosphere from the Lawrence Livermore National Laboratory can be found at https://cams.llnl.gov/cams-competencies/forensics/14c-bomb-pulse-forensics. The measurements showing the excess crossing zero in 2021 are from H. Graven et al., "Radiocarbon Dating: Going Back in Time," *Nature* 607 (2022): 449.

5. P. J. Reimer et al., "The INTCAL20 Northern Hemisphere Radiocarbon Age Calibration Curve (0–55 CAL kBP)," *Radiocarbon* 62, no. 4 (2020): 725–757.

6. Note that "years before the present (BP)" are defined by using 1950.0 as the "present" (when radiocarbon dating was first demonstrated) to avoid having to update graphs and tables constantly with the current date.

7. F. Miyake, K. Nagaya, K. Masuda, and T. Nakamura, "A Signature of Cosmic Ray Increase in AD 774–775 from Tree Rings in Japan," *Nature* 486 (2012): 240–242.

8. Retrieved from the Yale University online version of the *Anglo-Saxon Chronicle* at https://avalon.law.yale.edu/medieval/ang08.asp.

9. In counting experiments (be it voting polls or high-speed electrons from radioactive decay), the uncertainty in the result is given by the square root of the number of counts. Thus, for a 1 percent uncertainty, one needs $10,000 \text{ cts} \pm \text{sqrt}(10,000) = 10,000 \text{ cts} \pm 100 \text{ cts}$, giving a fractional error of $100/10,000 = 0.01 = 1\%$.

10. Given that the parchment is roughly 1,420 years old, the number of C-14 atoms remaining, N, will be (from p 80) the number it started with $\times (\frac{1}{2})^{1420/5730} = 0.86$. Half of those remaining atoms will decay in the next 5,730 years, which means the number decaying in the next eight-hour day in the lab is $N\{2 \times 5,730 \text{ year} \times 365 \text{ days/year} \times 3 \text{ 8-hour intervals per day}\}$ or $N/1.25 \times 10^7$. Recall that there are only 10^{-12} C-14 atoms for every C-12 atom; we want 10,000 counts in our eight-hour day (see endnote 9 above). This means that we need $10^4 \times 1.25 \times 10^7 \times 10^{12}/0.86 = 1.5 \times 10^{23}$ Carbon atoms in our sample. Because each Carbon atom has a mass of $12 \times 1.66 \times 10^{-24}$ g, and C makes up about 20 percent of the mass of the parchment, we need $5 \times 12 \times 1.66 \times 10^{-24} \text{ g/atom} \times 1.5 \times 10^{23} \text{ atoms} = 15 \text{ g of}$ parchment—the entire weight of several sheets of the manuscript.

11. An age of 35,000 years is just over six half-lives of C-14, meaning only $(\frac{1}{2})^6$ of the atoms (1.6 percent) in the original living object remain. But 0.01 g of Carbon contains $(0.01 \text{ g})/(12 \text{ amu/atom} \times 1.66 \times 10^{-24} \text{ g/amu}) = 5 \times 10^{20}$ atoms of Carbon, 10^{-12} of which were

originally the C-14 isotope, yielding 5×10^8 atoms of C-14. Now that only 1 percent remains, there are still 5 million C-14 atoms in the sample.

12. The uncertainty in a measurement is usually quoted as either one or two times the standard deviation of the measurement, a statistical description of the spread in repeated measurements of the same quantity. Plus or minus two standard deviations provides 95 percent confidence (95.4 percent, to be precise) that the true value lies in the quoted range of years, which means one standard deviation is $(645 - 568)/4 = 77/4 = 19$ years, and so ± 3 standard deviations (which gives 99 percent confidence) would be 549 to 663 years.

13. Travis Gettys, "Carbon Dating Suggests 'World's Oldest' Koran Could Be Older Than the Prophet Muhammad," *Raw Story*, August 31, 2015, https://www.rawstory.com/2015/08/carbon-dating-suggests-worlds-oldest-koran-is-even-older-than-the-prophet-muhammad/. Note that, while there are literally dozens of news stories about dating this Quran, including the original British Broadcasting Corporation (BBC) story from July 2015 (Sean Coughlan, " 'Oldest' Koran Fragments Found in Birmingham University," BBC, July 22, 2015, https://www.bbc.com/news/business-33436021) based on the University of Birmingham press release (University of Birmingham, "Birmingham Qur'an Manuscript Dated Among the Oldest in the World," press release, July 22, 2015, https://www.birmingham.ac.uk/news/latest/2015/07/quran-manuscript-22-07-15.aspx), I could find no publication of the result in a refereed scientific journal.

14. I. Hajdas, "Applications of Radiocarbon Dating Methods," *Radiocarbon* 51, no. 1 (2009): 79-90.

15. Much of this section is based on papers in a special Spring 1982 issue of *The Skeptical Inquirer* (vol. 6, no. 3), which is devoted to the Shroud of Turin.

16. M. Mueller, "The Shroud of Turin: A Critical Appraisal," *Skeptical Inquirer* 6, no. 3 (1982): 18.

17. Gordon Govier, "The Shroud's Second Image," *Christianity Today*, December 1, 2004, https://www.christianitytoday.com/ct/2004/december/32.56.htm.

9. HISTORY WITHOUT WORDS: LIME AND LEAD AND POOP

1. With a dollop of luck and some careful and creative techniques, a team of scientists and art historians cracked this problem as described in the following section, which is largely based on the article by J. Hale, J. Heinemeier, L. Lancasater, A. Lindroos, and A Rongbom "Dating Ancient Mortar," *American Scientist* 91 (2003): 130–137.

2. H. Delile, J. Blichert-Toft, J.-P. Goiran, S. Keay, and F. Albarede, "Lead in Ancient Rome's City Waters," *PNAS* 111, no. 18 (2014): 6594–6599.

3. See F. P. Retief and L. Cilliers, "Lead Poisoning in Ancient Rome," *Acta Theologica Supplementum* 7 (2005): 147, for an extensive disquisition on Lead in ancient Rome.

4. J. Russ, M. Hyman, H. J. Shafer, and M. Rowe, "Radiocarbon Dating of Prehistoric Rock Painting by Selective Oxidation of Organic Carbon," *Nature* 348 (1990): 710–711.

5. A. Quiles et al., "A High-Precision Chronological Model for the Decorated Upper Paleolithic Cave of Chauvet-Pont d'Arc, Ardèche, France," *PNAS* 113, no. 17 (2016): 4670–4675.

6.   P. Guibert et al., "When Were the Walls of the Chauvet-Pont d'Arc Cave Heated? A Chrono-logical Approach by Thermoluminescence," *Quaternary Geochronology* 29 (2015): 36–47.

7.   M. Aubert, A. Brumm, and J. Huntley, "Early Dates for 'Neanderthal Cave Art' May Be Wrong," *Journal of Human Evolution* 125 (2018): 215–217.

8.   M. Aubert et al., "Earliest Hunting Scene in Prehistoric Art," *Nature* 576 (2019): 442–445.

9.   Ed Whelan, "World's Oldest Stone Tools and Weapons Found in Ethiopia," Ancient-Origins.net, June 4, 2019, https://www.ancient-origins.net/news-history-archaeology /stone-tools-0012061.

10. From G. Wagner and P. van den Haute, *Fission Track Dating* (Dordrecht: Kluwer, 1992).

11. For a total fission energy of 170 MeV, each fragment carries away roughly half, that is, 85 MeV. The typical molecular bond in a mineral has an energy of approximately 5 eV, which means the fission fragment can break about 85×10^6 eV/5eV $= 1.7 \times 10^7$ molecules. If each molecule is approximately 0.5 nm in size, the fragment can travel 1.7×10^7 mole-cules $\times 0.5 \times 10^{-9}$ m/molecule of approximately 8.5×10^{-3} m or 8 mm.

12. M. J. Morwood, P. B. O'Sullivan, F. Aziz, and A. Raza, "Fission-Track Ages of Stone Tools and Fossils on the East Indonesian Island of Flores," *Nature* 392 (1998): 173.

13. J.-J. Hublin et al., "New Fossils from Jebel Irhoud Morocco and the Pan-African Origin of Home Sapiens," *Nature* 546 (2017): 289–292.

14. J.-J. Hublin, N. Sirakov, and T. Tsenka, "Initial Upper Paleolithic Homo Sapiens from Bacho Kiro Cave, Bulgaria," *Nature* 581 (2020): 299–302.

15. G. A. Person and J. W. Ream, "Clovis on the Caribbean Coast of Venezuela," *Current Research in the Pleistocene* 22 (2005): 28–31.

16. D. L. Jenkins et al., "Clovis-Age Western Stemmed Projectile Points and Human Copro-lites at the Paisley Caves," *Science* 337 (2012): 223–228.

17. L.-M. Shillito et al., "Pre-Clovis Occupation of the Americas Identified by Human Fecal Biomarkers in Coprolites from Paisley Caves, Oregon," *Science Advances* 6 (2020): eaba6404.

18. A. J. Lesnek, J. P. Briner, C. Lindqvist, J. F. Baichtal, and T. H. Heaton, "Deglaciation of the Pacific Coast Corridor Directly Preceded the Human Colonization of the Americas," *Science Advances* 4, no. 5 (2018): eaar5040.

10. YOU ARE WHAT YOU EAT

1.   R. B. Richardson, D. S. Allan, and Y. Le, "Greater Organ Involution in Highly Prolifer-ative Tissues Associated with the Early Onset and Acceleration of Ageing in Humans," *Experimental Gerontology* 55 (2014): 80–91.

2.   This estimate is for a 70-kg adult male. If you are having trouble picturing 3000 trillion trillion, imagine a warehouse that covers the entire surface of the Earth (land and sea) and is as tall as the Empire State Building, full of poppy seedss. That's 3000 trillion tril-lion poppy seeds.

3.   N. J. van der Merve, "Carbon Isotopes, Photosynthesis, and Archeology," *American Scien-tist* 70 (1982): 596–606.

4.   R. J. Forbes, "Metallurgy in Antiquity," (Leiden: E.J. Brill, 1950), as cited in Nikolaas J. van der Merwe and M. Stuiver, "Dating Iron by the C-14 Method," *Current Anthropology* 9, no. 1 (1968): 48–53.

5. The 2018–2019 estimates for worldwide grain production are found at M. Shahbandeh, "Worldwide Production of Grain in 2022/23, by Type (in Million Metric Tons)," Statista, February 16, 2023, https://www.statista.com/statistics/263977/world-grain-production -by-type/, and are reproduced in the following table:

Grain	Annual production (Mtons)	Percent of total
Corn*	1,100 million metric tons	42.5 percent
Wheat	735	
Rice	496	
Barley	141	
Sorghum*	58.4	2.25 percent
Oats	22.2	
Rye	10.6	
Millet*	27.8	1.07 percent

*C4 plants total = 46 percent

6. The technical standard for the $^{13}C/^{12}C$ ratio, as arbitrary as is any standard like the meter or the kilogram, is a fossil seashell from South Carolina. Compared to this shell, air had a ratio of −0.7 percent when this research was done in the 1980s. To avoid adding this complication here, I have just used the value of air as the standard against which all other ratios are compared. To compare my numbers to the scientific literature, however, one must add −0.7 percent to all values quoted here.

7. Although we shan't discuss them further, for the record, CAM plants have a ratio close to that of C4 plants. See R. H. Tykot, "Stable Isotopes and Diet: You Are What You Eat," in *Proceedings of the International School of Physics "Enrico Fermi" Course CLIV*, ed. M. Martini, M. Milazzo, and M. Piacentini (Amsterdam: IOS Press, 2004), 435.

8. Bone is about 18 percent Carbon, so 1 g of bone contains $1g/(12 \text{ amu/atom} \times 1.67 \times 10^{-24}$ g/amu$) \times 0.18 = 1.6 \times 10^{21}$ atoms of Carbon, of which 1.2 percent is C-13 and the discrepancy between the C3 and C4 diets is 1.4 percent of that, so $9 \times 10^{21} \times 0.012 \times 0.014 = 1.5 \times 10^{18}$ atoms or 1.5 million trillion atoms.

9. D. R. Piperno and K. V. Glannery, "The Earliest Archeological Maize (Zea mays L.) from Highland Mexico: New Accelerator Mass Spectrometry Dates and Their Implications," *PNAS* 98, no. 4 (2001): 2101–2103.

10. Shahbandeh, "Worldwide Production of Grain in 2022/23," https://www.statista.com /statistics/263977/world-grain-production-by-type/.

11. The combination of food with a carbon isotope ratio of −1.45 (C3 plants) × 30% + a ratio of 0.0 (C4 plants) × 70% = the observed value of −0.45 percent.

12. A. C. Roosevelt, "The Development of Prehistoric Complex Societies: Amazonia, a Tropical Forest," in *Complex Polities in the Ancient Tropical World*, ed. E. A. Bacus, L. J. Lucero, and J. Allen (Arlington, VA: American Anthropological Association, 1999), 13–34.

13. E. Medina and P. E. H. Minchin, "Stratification of ^{13}C in Amazonian Rainforests," *Oecologia* 45 (1980): 337–378.

14. This is not the only way in which air can depart from its normal Carbon isotope ratio. For example, the ratio in Los Angeles air is −0.2 percent below the standard value because of the constant addition of automobile exhaust that, of course, comes from burning gasoline, ^{13}C-depleted plant material from 200 million years ago. In fact, the global value of ^{13}C/^{12}C has fallen from −0.64 percent compared to the standard in preindustrial times to −0.86 percent today as a consequence of fossil fuel combustion.

15. *Quod erat demonstrandum* is the Latin phrase, usually abbreviated QED that follows a mathematical proof and means "which was to be shown," that is, "we have demonstrated what we set out to prove."

16. Some members of the Archaean kingdom, the most primitive form of life still extant on Earth, also fix Nitrogen, and are especially important in Oxygen-poor soils where bacteria cannot survive.

17. N. A. Campbell and J. B. Reece, "Biology," (San Francisco: Pearson Benjamin Cummings, 2005).

18. M. J. Schoeninger, M. J. DeNiro, and H. Tauber, "Stable Nitrogen Isotope Ratios of Bone Collagen Reflect Marine and Terrestrial Components of Prehistoric Human Diet," *Science* 220, no. 4604 (1983): 1381–1383.

19. B. Buchardt, V., Bunch V., and P. Helin, "Fingernails and Diet: Stable Isotope Signatures of a Marine Hunting Community from Modem Uummannaq, North Greenland," *Chemical Geology* 244 (2007): 316–329.

20. K. A. Hobson, R. T. Alisauskas, and R. G. Clark, "Stable-Nitrogen Isotope Enrichment in Avian Tissues Due to Fasting and Nutritional Stress: Implications for Isotopic Analyses of Diet," *The Condor* 95, no. 2 (1993): 388.

21. L. J. Reitsema, "Beyond Diet Reconstruction: Stable Isotope Applications to Human Physiology, Health, and Nutrition," *American Journal of Human Biology* 25 (2013): 445–456.

22. C. M. Cook, A. L. Alvig, Y. Q. Liu, and D. A. Schoeller, "The Natural ^{13}C Abundance of Plasma Glucose is a Useful Biomarker of Recent Dietary Caloric Sweetener Intake," *Journal of Nutrition* 140, no. 2 (2010): 333–337.

23. P. S. Patel et al., "Serum Carbon and Nitrogen Stable Isotopes as Potential Biomarkers of Dietary Intake and Their Relation with Incident Type 2 Diabetes: The EPIC-Norfolk Study," *American Journal of Clinical Nutrition* 100 (2014): 708–718.

24. K. J. Petzke, T. Feist, W. E. Fleig, and C. C. Metges, "Nitrogen Isotopic Composition in Hair Protein is Different in Liver Cirrhotic Patients," *Rapid Communications in Mass Spectrometry* 20, no. 19 (2006): 2973–2978.

25. T.-C. Kuo et al., "Assessment of Renal Function by the Stable Oxygen and Hydrogen Isotopes in Human Blood Plasma," *PLOS ONE* 7, no. 2 (2012): e32137.

26. R. Prinoth-Fornwagner and T. R. Niklaus, "The Man in the Ice: Results from Radiocarbon Dating," *Nuclear Instruments and Methods in Physics Research Section B: Beam Interactions with Materials and Atoms* 92, no. 1–4 (1994): 282–290.

27. W. Muller, H. Fricke, S. A. N. Halliday, M. T. McCulloch, and J. Wartho, "Origin and Migration of the Alpine Iceman," *Science* 302, no. 5646 (2003): 862–866.

28. M. J. Wooller, C. Bataille, P. Druckenmeir, G. M. Erickson, P. Groves, N. Haubenstock, T. Howe, J. Irrgeher, D. Mannm, and A. D. Willis, "Lifetime Mobility of an Arctic Woolly Mammoth," *Science* 373, no. 6556 (2021): 806–808.

29.   J. H. Miller, D. C. Fisher, B. E. Crowley, and B. A. Konomi, "Male Mastodon Landscape Use Changed with Maturation," *PNAS* 119, no. 25 (2022): e2118329119.

11. PALEOCLIMATE: TAKING THE EARTH'S TEMPERATURE LONG AGO

1.   Wally Broecker transferred to Columbia College from Wheaton College in Illinois to complete his undergraduate degree in 1953 after a summer research experience at Columbia's Lamont-Doherty Earth Observatory. He remained with the university for the next sixty-six years, until his death in February 2019. He authored more than 500 journal articles on geochronology, radiocarbon dating, and chemical oceanography. He proposed the idea of the "ocean conveyor" charting the currents that carry energy around the world. He made seminal contributions to climate change science and spoke out for forty-five years on the dangers of global warming, characterizing the climate as "an angry beast" that we would do well not to provoke. Broecker was winner of the Crafoord Prize and the Vetlesen Prize and was awarded the National Medal of Science by President Bill Clinton in 1996. See his Wikipedia entry and the following page for further details of his extraordinary and inspirational career: Earth Institute, Columbia University, https://www.earth.columbia.edu/articles/view/2246.

2.   W. S. Broecker, "Climatic Change: Are We on the Brink of a Pronounced Global Warming?," *Science* 189 (1975): 460–463.

3.   Broecker, "Climatic Change."

4.   The remaining 0.028 percent comes from sunlight reflected by the Moon and other planets (0.00013 percent), light from other stars (less than 0.0001 percent), tidal friction caused by the Moon (0.0019 percent), heat leaking from below Earth's surface owing to the initial heat of formation of the planet and the radioactive decay of long-lived isotopes in the Earth's crust (0.026 percent, the dominant other component), and matter falling to Earth (which, with the rare exception of asteroid collisions—see chapter 12) is less than 0.000001 percent.

5.   It is not an accident that our eyes are sensitive to the single octave of light that matches the Sun's maximum output; this is a product of evolution through natural selection. Our light receptors have been tuned over 400 million years of evolution to take advantage of the most abundant light source present on the surface of the Earth, which is simply a function of the Sun's temperature and the transparency of the Earth's atmosphere at these wavelengths. Had we evolved on a star with a different temperature or on a planet with a different atmospheric composition, we should expect to have light sensors tuned to different wavelengths.

6.   Charles David Keeling developed the first instrument capable of reliably measuring the concentration of carbon dioxide (CO_2) in the atmosphere while he was a postdoctoral fellow at the California Institute of Technology in the 1950s. In 1956, he joined the research staff at the Scripps Institution of Oceanography, and in 1958, he deployed his instrument on Mauna Loa in Hawaii, where updated versions continue to measure the CO_2 concentration daily under the supervision of one of his sons, Ralph Keeling, also a professor at Scripps.

7.   For up-to-the-minute measurements, see "Daily CO_2," Co_2.Earth, www.co$_2$.earth/daily-co$_2$.

8. J. R. Dean, M. J. Leng, and A. W. Mackay, "Is There an Isotopic Signature of the Anthropocene?," *Anthropocene Review* 1, no. 3 (2014): 276–287.

9. H. Graven, "Impact of Fossil Fuel Emissions on Atmospheric Radiocarbon and Various Applications of Radiocarbon Over This Century," *PNAS* 112, no. 31 (2015): 9542–9545.

10. Because Austria is a landlocked country, it might seem odd to name the isotopic standard for ocean water after Vienna. But the VMSOW was defined in 1968 by the International Atomic Energy Agency, which has its headquarters in Vienna. VMSOW is now supported by the U.S. National Institute of Standards and Technology and the European Institute for Reference Materials and Measurements.

11. The molecule $H_2^{18}O$ has a mass of 20 atomic mass units, or 11.1 percent higher the $H_2^{16}O$, whereas $^2H^1HO$ has a mass of 19 amu, so it is only 5.6 percent heavier.

12. M. F. Porter, J. Pisaric, S. V. Kokelj, and T. W. D. Edwards, "Climatic Signals in $\delta^{13}C$ and $\delta^{18}O$ of Tree-rings from White Spruce in the Mackenzie Delta Region, Northern Canada," *Arctic, Antarctic, and Alpine Research* 41, no. 4 (2009): 497–505.

13. T. W. D. Edwards et al., "^{13}C Response Surface Resolves Humidity and Temperature Signals in Trees," *Geochimica et Cosmochimica Acta* 64, no. 2 (2000): 161–167.

14. C. Loehle, "Correction to a 2,000-Year Global Temperature Reconstruction Based on Non-Tree Ring Proxies," *Energy and Environment* 18, no. 7 (2007): 1049–1058.

15. Loehle, "Correction to a 2,000-Year Global Temperature Reconstruction."

16. L. L. Dorman, "Chapter 30—Space Weather and Cosmic Ray Effects," in *Climate Change*, 2nd ed., ed. T. M. Letcher (Amsterdam: Elsevier, 2016), 513–544.

17. C. A. Woodhouse, D. M. Meko, G. M. MacDonald, D. W. Stahle, and E. R. Cook, "A 1,200-Year Perspective of 21st Century Drought in Southwestern North America," *PNAS* 107, no. 50 (2010): 21283–21288.

18. J. Esper, "Orbital Forcing of Tree-ring Data," *Nature Climate Change* 2 (2012): 862–866.

19. L. Loulergue, "Orbital and Millennial-Scale Features of Atmospheric CH_4 over the Past 800,000 Years," *Nature* 453 (2008): 383–386.

20. A. E. Carlson, "Ice Sheets and Sea Level in Earth's Past," *Nature Education Knowledge* 3, no. 10 (2011): 3, figure 3.

21. Loulergue, "Orbital and Millennial-Scale Features of Atmospheric CH_4."

22. Many versions of this plot exist; an example can be found in Ngai Weng Chan, Seow Wee, David Martin, Kai Chen Goh, and Hui Hwang Goh, "Global Warming," in *Sustainable Urban Development Textbook*, ed. Ngai Weng Chan, Hidefumi Imura, Akihiro Nakamura, and Masazumi Ao (Water Watch Penang & Yokohama City University), 67–73.

23. Rebecca Lindsey and Luann Dahlman, "Climage Change: Ocean Heat Content," Climate .gov, August 17, 2020, https://www.climate.gov/news-features/understanding-climate /climate-change-ocean-heat-content.

24. Nick Bradford, "A Warming Ocean," National Environmental Education Center, n.d., accessed February 2, 2023, https://www.neefusa.org/nature/water/warming-ocean and references therein.

25. You can find similar stories via an internet search engine, such as https://www.bing .com/search?q=news-features/climate-and/climate-lobsters&FORM=ATUR01&PC =ATUR&PTAG=ATUR01RAND.

26.   In February 2015, following a snowstorm in Washington, DC, Senator James Inhofe, Republican from Oklahoma, (R-Oklahoma) made a snowball outside the Capitol, brought it into the Senate chamber, and threw it; it was his absurd attempt to support his theory that human-induced climate change is, to quote the title of his book, *The Greatest Hoax*. See, for example, Timothy Cama, "Inhofe Hurls Snowball on Senate Floor," *The Hill*, February 26, 2015, https://thehill.com/policy/energy-environment/234026-sen-inhofe-throws -snowball-to-disprove-climate-change

27.   Hannah Bailey et al., "Arctic Sea-Ice Loss Fuels Extreme European Snowfalls," *Nature Geoscience* 14 (2021): 283–288.

28.   M. Sigl, "Timing and Climate Forcing of Volcanic Eruptions for the Past 2,500 Years," *Nature* 523 (2015): 543–549.

29.   F. Lavigne et al., "Source of the Great A.D. 1257 Mystery Eruption Unveiled, Samalas Volcano, Rinjani Volcanic Complex, Indonesia," *PNAS* 110, no. 42 (2013): 16742–16747.

30.   Richard B. Stothers, "Climatic and Demographic Consequences of the Massive Volcanic Eruption of 1258," *Climatic Change* 45 (2000): 361–374

31.   A. V. Kurbatov et al., "A 12,000-Year Record of Explosive Volcanism in the Siple Dome Ice Core, West Antarctica," *Journal of Geophysical Research* 111 (2006): D12307.

32.   Sigl, "Timing and Climate Forcing of Volcanic Eruptions."

33.   F. Mikhaldi et al., "Multiradionuclide Evidence for the Solar Origin of the Cosmic-Ray Events of AD 774/5 and 993/4," *Nature Communications* 6, no. 8611 (2015).

34.   Tony Phillips, "Severe Space Weather--Social and Economic Impacts," Nasa Science, January 21, 2009, https://science.nasa.gov/science-news/science-at-nasa/2009/21jan_severe spaceweather/.

35.   Auroras are rarely seen at all as far south as New England, let alone appearing bright enough to light up the sky; they have never been seen south of the Mason Dixon line except for this event.

36.   Lloyds, with Atmospheric and Environmental Research (AER) *Solar Storm Risk to the North American Electric Grid*, May 2013, https://www.lloyds.com/news-and-insights/risk -reports/library/solar-storm.

37.   As light, cosmic rays, or whatever leaves an object, it or they form a spherical shell around the source. The surface area of this sphere grows as its radius squared (R^2); thus, when the sphere has expanded tenfold, its impact is 100 times less over each square meter of an object it hits (for light, the source would appear only 1 percent as bright). Therefore, for a supernova thirty times closer, the effect would be 30^2, or 900 times as great.

38.   B. Schwarzschild "Recent Nearby Supernovae May Have Left Their Marks on Earth," *Physics Today* 55, no. 5 (2002): 19.

39.   D. Koll et al., "Interstellar ^{60}Fe in Antarctica," *Physical Review Letters* 123 (2019): 072701–1-6.

40.   Calcium carbonate shells of foraminifera and coccoliths (their plant counterparts), and the silicon dioxide shells of radiolarians (animals) and diatoms (tiny plants) all contain oxygen.

41.   Lake sediments can be used similarly.

42.   Y. Morono, "Aerobic Microbial Life Persists in Oxic Marine Sediment as Old as 101.5 Million Years," *Nature Communications* 1, no. 3626 (2020) 1–9.

12. THE DEATH OF THE DINOSAURS: AN ATOMIC VIEW

1. Genera is the plural of genus. Genus is one step above a species in the classification of organisms and describes a group of closely related, similar species.

2. J. Phillips, *Life on Earth: Its Origins and Successions* (Cambridge: MacMillan, 1860), 58–59.

3. G. Ceballos, P. R. Ehrlich, and R. Dirzo, "Biological Annihilation via the Ongoing Sixth Mass Extinction Signaled by Vertebrate Population Losses and Declines," *PNAS* 114, no. 30 (2017): E6089–E6096; E. Kolbert, *The Sixth Extinction: An Unnatural History* (New York: Henry Holt, 2014).

4. This used to be called the "K-T boundary," where the "K" stood for the German translation of Cretaceous, and the "T" stood for Tertiary. In 2008, the Tertiary was divided into the Paleogene and the Neogene by the International Commission on Stratigraphy; the Paleogene begins 66 Myr ago.

5. L. W. Alvarez, W., Alvarez, F. Asaro, and H. V. Michel, "Extraterrestrial Cause for the Cretaceous-Tertiary Extinction," *Science* 208, no. 4448 (1980): 1095–1108.

6. Gravitational anomalies are departures from the gravitational pull one would expect at a given location on the Earth if the Earth were a perfectly smooth sphere. Magnetic anomalies are discrepancies from what one would expect if the Earth's magnetic field were a smooth and perfect dipole (magnet with equally strong north and south poles). Both kinds of anomalies occur at the location of an impact crater.

7. *Chicxulub* is a Mayan word meaning "the devil's flea" and is the name of the pueblo (village) closest to the center of the crater. V. R. Bricker, E. O. Yah, and O. D. de Po'ot, *A Dictionary of the Mayan Language* (Salt Lake City: University of Utah Press, 1998).

8. T. E. Krogh, S. L. Kamo, and B. F. Bohor, "Fingerprinting the K/T Impact Site and Determining the Time of Impact by UPb Dating of Single Shocked Zircons from Distal Ejecta," *Earth and Planetary Science Letters* 199 (1993): 425–429.

9. Argon has three stable isotopes, ^{36}Ar, ^{38}Ar, and ^{40}Ar, but the latter makes up 99.6 percent of the total.

10. P. R. Renne et al., "Timescales of Critical Events Around the Cretaceous-Paleogene Boundary," *Science* 339 (2013): 648.

11. J. A. Wolfe, "Paleobotanical Evidence for the June 'Impact Winter' at the Cretaceous/Tertiary Boundary," *Nature* 352 (1991): 420.

12. J. Vellekoop et al. "Rapid Short-Term Cooling Following the Chicxulub Impact at the Cretaceous–Paleogene Boundary," *PNAS* 111, no. 21 (2014): 7537–7541.

13. B. K. Nelson, G. K. MacLeod, and P. D. Ward, "Rapid Change in Strontium Isotopic Composition of Sea Water Before the Cretaceous/Tertiary Boundary," *Nature* 351 (1991): 644–647.

14. J. Hess, M. L. Bender, and J.-G. Schilling, "Evolution of the Ratio of Strontium-87 to Strontium-86 in Seawater from Cretaceous to Present," *Science* 231, no. 4741 (1986): 979–984.

15. J. D. MacDougall, "Seawater Strontium Isotopes, Acid Rain, and the Cretaceous-Tertiary Boundary," *Science* 239, no. 4839 (1987): 485–487.

16.   J. Smit, "Cretaceous-Tertiary Boundary Impact Ejecta," *Annual Review of Earth and Planetary Sciences* 27 (1999): 75–113.

17.   W. S. Wolbach, R. S. Lewis, and E. Anders, "Cretaceous Extinctions: Wildfires and the Search for Meteoritic Material," *Science* 230, no. 4722 (1985): 167–170.

18.   J. Morgan, N. Artemieva, and T. Goldin, "Revisiting Wildfires at the K-Pg Boundary," *Journal of Geophysics Research Biosciences* 118, no. 4 (2013): 1508–1520.

19.   R. A. DePalma et al., "A Seismically Induced Onshore Surge Deposit at the K-Pg Boundary, North Dakota," *PNAS* 116, no. 17 (2019): 8190–8199.

20.   DePalma et al., "A Seismically Induced Onshore Surge Deposit."

13. EVOLUTION: FROM METEORITES TO CYANOBACTERIA

1.   Israel Science and Technology homepage at https://www.science.co.il/elements/?s=Earth.

2.   This is simply a result of the fact that, to identify a molecule, we need to know its spectrum; lots of other spectral lines belong to yet-to-be-identified molecules.

3.   Wikipedia, s.v. "List of Interstellar and Circumstellar Molecules," last updated April 28, 2023, https://en.wikipedia.org/wiki/List_of_interstellar_and_circumstellar_molecules.

4.   S. A. Wilde, J. W. Valley, W. H. Peck, and C. M. Graham, "Evidence from Detrital Zircons for the Existence of Continental Crust and Oceans on the Earth 4.4 Gyr Ago," *Nature* 409 (2001): 175–177.

5.   J. O. O'Neil, J.-L. Paquette, J.-L. and D. Francis, "Formation Age and Metamorphic History of the Nuvvuagittuq Greenstone Belt," *Precambrian Research* 220–221 (2012): 23–44.

6.   T. Tashiro et al., "Early Trace of Life from 3.95 Ga Sedimentary Rocks in Labrador, Canada," *Nature* 549 (2017): 516–518.

7.   W. Lenz, "A Short History of Thalidomide Embryopathy," *Teratology* 38 (1988): 203–213.

8.   Actually, the story is somewhat more complicated. See E. Tokunaga, *Scientific Reports* 8 (2018): 17131.

9.   Two more are found in a few species of bacteria and archaea.

10.   S. H. Gellman and D. N. Woolfson, "Designing a 20-Residue Protein," *Nature Structural Biology* 9, no. 6 (2002): 425–430; and K. H. Mok et al., "A Pre-existing Hydrophobic Collapse in the Unfolded State of an Ultrafast Folding Protein," *Nature* 447 (2007): .7140.

11.   J. E. Elsila et al., "Meteoritic Amino Acids: Diversity in Compositions Reflect Parent Body Histories," *ACS Central Science* 2, no. 6 (2016): 370–379.

12.   J. Sokol, "An Unusual Meteorite, More Valuable Than Gold, May Hold the Building Blocks of Life," *Science* (2020), doi:10.1126/science.abe.3025.

13.   D. P Glavin et al., "Unusual Non-terrestrial L-proteinogenic Amino Acid Excesses in the Tagish Lake Meteorite," *Meteoritics & Planetary Science* 47 (2012): 1347–1364.

14.   T. Lee and Y. K. Lin, "The Origin of the Life and the Crystallization of Aspartic Acid in Water," *Crystal Growth and Design* 10 (2010): 1652–1660.

15.   A. Dylewski, "Meteorites Delivered the "Seeds" of Earth's Left-Hand Life," American Chemical Society, press release, April 6, 2008, https://www.acs.org/content/acs/en/pressroom /newsreleases/2008/april/meteorites-delivered-the-seeds-of-earths-left-hand-life.html; and M. Levine, C. S. Kenesky, D. Mazori, and R. Breslow, "Enantioselective Synthesis

and Enantiomeric Amplification of Amino Acids Under Prebiotic Conditions," *Organic Letters* 10 (2008): 2433–2436.

16. It should be noted that the sugars we can digest are right-handed.

17. D. P. Glavin and J. P. Dworkin, "Enrichment of the Amino Acid l-Isovaline by Aqueous Alteration on CI and CM Meteorite Parent Bodies," *PNAS* 106, no. 14 (2009): 5487–5492.

18. M. Peplow, "Meteorite Molecules Spin Sugars," *Nature* (2004), https://doi.org/10.1038 /news040216-18.

19. Y. Furukawaa et al., "Extraterrestrial Ribose and Other Sugars in Primitive Meteorites," *PNAS* 116 (2019): 24440–24445.

20. Also sometimes called the last universal common ancestor (LUCA).

14. WHAT'S UP IN THE AIR? EARTH'S EVOLVING ATMOSPHERE

1. The *Voyager 1* spacecraft, launched in 1977 carrying the famous gold record with humanity's greetings to the cosmos, crossed the boundary between the solar system and interstellar space around 2012. *Voyager 2* crossed that boundary in November 2018, when it was 17.8 billion km from Earth.

2. D. C. Catling, and K. J. Zahnle, "The Escape of Planetary Atmospheres," *Scientific American* 300, no. 5 (2009): 36–43.

3. Banded iron formations are thought to have originated from seawater in which the first photosynthetic organisms produced Oxygen, which then combined with Iron in the water to make iron oxide on the sea floor.

4. A. J. Charles et al., "Constraints on the Numerical Age of the Paleocene-Eocene Boundary," *Geochemistry, Geophysics, Geosystems* 12, no. 6 (2011):1–19

5. M. Guthar et al., "Very Large Release of Mostly Volcanic Carbon During the Palaeocene–Eocene Thermal Maximum," *Nature* 548, no. 7669 (2017): 573–577.

6. A massive (6.6 million cubic kilometers) flood of magma that now makes up Iceland and parts of Norway, Scotland, Ireland, and Greenland.

7. G. H. Denton et al., "The Last Glacial Termination," *Science* 328, no. 5986 (2010): 1652–1656.

8. J. F. McManus, R. Francois, J.-M. Gherardi, L. D. Keigwin, and S. Brown-Leger, "Collapse and Rapid Resumption of Atlantic Meridional Circulation Linked to Deglacial Climate Changes," *Nature* 428 (2004): 834–837.

9. R. F. Anderson, "Wind-Driven Upwelling in the Southern Ocean and the Deglacial Rise in Atmospheric CO_2," *Science* 323, no. 5920 (2009): 1443–1448.

10. The Younger Dryas interval of rapid climate change is named after the tundra wildflower *Dryas octopetala*, the leaves of which are frequently found in Scandinavian lake sediments from this period. The recent dating comes from an analysis of trees buried in one of Europe's largest volcanic eruptions, the Laacher See, which brought into agreement lake sediment and ice core records and allowed us to date the onset of the Younger Dryas with an astounding precision of better than 0.1 percent (equivalent to looking at me and guessing my birthdate to within a month). See R. Reinig et al., "Precise Date for the Laacher See Eruptions Synchronizes the Younger Dryas," *Nature* 595 (2021): 66–69.

11.   K. Andreassen et al., "Massive Blow-out Craters Formed by Hydrate-Controlled Methane Expulsion from the Arctic Seafloor," *Science* 356 (2017): 948–953.

12.   D. D. Catling and K. J. Zahnle, "The Escape of Planetary Atmospheres," *Scientific American* 300, no. 5 (2009): 36–43.

13.   Contrary to popular belief, the ozone hole problem is unrelated to global warming. Chemicals containing Chlorine, such as refrigerants that use chlorofluorocarbon compounds, drift into the stratosphere and there destroy ozone molecules, allowing more UV light to penetrate through the atmosphere. While this can be detrimental to ocean-dwelling and land-dwelling life, all of which has evolved under the normally protective ozone blanket, it does not affect the Earth's temperature. Ground-level (tropospheric) ozone is a greenhouse gas, however, and adds almost 10 percent to the warming effect of CO_2, H_2O, methane, and other such gases.

15. OUR SUN'S BIRTHDAY: THE SOLAR SYSTEM IN FORMATION

1.   To a physicist, this is known as conservation of angular momentum; for an object moving in a circle, the quantity mass × velocity × radius of the circle remains constant unless acted upon by an outside force.

2.   The principal component of all matter in the universe is dark matter, whose identity remains a mystery. There is about seven times as much mass in dark matter as there is in normal matter in the universe. But it appears that dark matter is at best weakly interacting; that is, it does not undergo collisions with normal matter or with itself. As a result, it remains in a spherical halo within which the flattened disk forms.

3. $$^{87}\mathrm{Rb}(t) = {}^{87}\mathrm{Rb}(t=0) \times e^{-0.693\,t/T} \quad \text{where } t = \text{times};\ T = \text{half-life}$$

Using today's measured values of $^{87}\mathrm{Rb}$, $^{87}\mathrm{Sr}$, and $^{86}\mathrm{Sr}$ in a sample leads directly to the age of the sample by measuring the slope of the line describing the isotopic ratios: "e" is the base of the natural logarithm and is equal to $2.718 - e^{-0.693\,t/T_{\frac{1}{2}}} = (\frac{1}{2})^{t/T_{\frac{1}{2}}}$ as used above.

Equation 1 is just the standard radioactive decay equation that says the amount of $^{87}\mathrm{Rb}$ present at any time t is the amount you started with (at $t=0$) minus the amount that has decayed.

$$^{87}\mathrm{Rb}(t=0) = {}^{87}\mathrm{Rb}(t) \times e^{+0.693\,t/T} \tag{1}$$

Equation 2 just rearranges equation 1 in a more convenient form.

$$^{87}\mathrm{Sr}(t) = {}^{87}\mathrm{Sr}(t=0) + \text{amount that was added by the decaying } {}^{87}\mathrm{Rb} \tag{2}$$

Equation 3 says that the amount of $^{87}\mathrm{Sr}$ you have now is the amount you started with plus the amount that has been added by the decaying $^{87}\mathrm{Rb}$, and that latter amount is just the amount of $^{87}\mathrm{Rb}$ you started with minus the amount now left (the rest having already decayed—see equation 4).

$$^{87}\mathrm{Sr}(t) = {}^{87}\mathrm{Sr}(t=0) + [{}^{87}\mathrm{Rb}(t=0) - {}^{87}\mathrm{Rb}(t)] \tag{3}$$

$$^{87}\mathrm{Sr}(t) = {}^{87}\mathrm{Sr}(t=0) + {}^{87}\mathrm{Rb}(t) \times e^{+0.693\,t/T} - {}^{87}\mathrm{Rb}(t) \tag{4}$$

Now, substituting equation 2 into equation 4, we get equation 5 for the amount of $^{87}\mathrm{Sr}$ we have in the sample right now.

$$\frac{^{87}\text{Sr}(t)}{^{86}\text{Sr}(t)} = \frac{^{87}\text{Sr}(t=0)}{^{86}\text{Sr}(t)} + \frac{\boxed{^{87}\text{Rb}(t)} \times e^{+0.693\,t/T}}{^{86}\text{Sr}(t)} - \frac{\boxed{^{87}\text{Rb}(t)}}{^{86}\text{Sr}(t)} \quad\longleftarrow \text{factor this out}$$
(5)

We then divide both sides of this equation by the amount of the stable isotope, ^{86}Sr, and rearrange terms to get equation 6.

$$\underset{\substack{\text{plot} \quad\; y\text{-intercept}}}{\underset{\substack{Y \;=\; \qquad b}}{\frac{\overset{\substack{\text{Measured}\\\text{today}}}{^{87}\text{Sr}(t)}}{^{86}\text{Sr}(t)} = \frac{^{87}\text{Sr}(t=0)}{^{86}\text{Sr}(t)}}} + \underset{\substack{\text{plot} \quad\text{line slope}\\ +\quad X \qquad m}}{\frac{\overset{\substack{\text{Measured}\\\text{today}}}{^{87}\text{Rb}(t)}}{^{86}\text{Sr}(t)} \times [e^{+0.693\,t/T} - 1]}$$
(6)

with labels: Measured today, Measured today, AGE!!, known half-life; plot y-intercept, plot line slope; Y = b + X m

Most conveniently, equation 6 is of the form:

$y = b + x \times m$ or $y = mx + b$

which (I am sure you all remember) is the simple equation of a straight line on an *x-y* graph, where *b* is the point at which the line intercepts the *y*-axis (i.e., the value of *y* at $x = 0$) and *m* is the slope of the line.

The quantity *y* here is the left-hand side of the equation, the ratio $^{87}\text{Sr}(t)/^{86}\text{Sr}(t)$, which is easily measured—it is just the number of atoms of each isotope that the sample contains right now. Likewise, the *x* term is $^{87}\text{Rb}(t)/^{86}\text{Sr}(t)$, the ratio of those two isotopes today. The *b* term, $^{87}\text{Sr}(t=0)/^{86}\text{Sr}(t)$, is the *y*-intercept and you can directly read off the graph the original amount of ^{87}Sr in the sample. Finally, one simply measures the slope of the line, sets it equal to $[e^{+0.693t/T} - 1]$ and plugs in the known half-life to find *t*, the age.

4.   J. N. Connelly et al. "The Absolute Chronology and Thermal Processing of Solids in the Solar Protoplanetary Disk," *Science* 338, no. 6107 (2012): 651–655.

5.   E. Gaidos, A. N. Krot, J. P. Williams, and S. N. Raymond, "^{26}Al and the Formation of the Solar System from a Molecular Cloud Contaminated by Wolf-Rayet Winds," *Astrophysical Journal* 696 (2009): 1854–1863.

16. STARDUST CREATION: BUILDING THE BUILDING BLOCKS

1.   The Moon and Sun appear to be *almost* the same size every day; small changes occur because of their elliptical orbits, which periodically change the distance between Earth and these bodies by a few percentage points.

2.   For comparison, the Earth has a mean density of 5.5 g/cm³.

3.   This very high temperature is required because of a quirk in the isotopes we discussed briefly in chapter 6. The only light element whose most stable nucleus doesn't have an equal number of protons and neutrons is Beryllium. ^{8}Be has a half-life of only 8.2×10^{-8} seconds and thus cannot serve as the next obvious step in the fusion process in which $^4\text{He} + {}^4\text{He} \rightarrow {}^8\text{Be}$ (rather, this does happen, but it immediately takes back the energy produced as it decays). ^{9}Be is the stable isotope, and there is no atomic mass 5 isotope to fuse with Helium to make it. Thus, we have to wait until the temperature is high

enough that the repulsive force of the twelve positive charges of three ^{4}He nuclei can be overcome by their high velocities and they get close enough to merge in ^{12}C, united by the strong force.

4. One might think that the more fuel a star has, the longer it will live, but in fact the opposite is the case: one must also consider the rate at which it uses up fuel. More massive stars generate higher core temperatures and densities, and because the nuclear reaction rate is exquisitely sensitive to temperature, more massive stars burn their supply of fuel faster. A star like the Sun has a total life span of nearly 12 *billion* years, while one with twenty-five times as much mass at birth only lives 7 *million* years.

5. NuSTAR is a satellite, designed and built at the California Institute of Technology and Columbia University. It is the first mission to image the sky in high-energy X-rays and thus the first to directly detect freshly synthesized radioactive elements in supernova remnants, including the remnant of the explosion observed in a nearby galaxy in 1987. See S. E. Boggs et al., *Science*, 348 (2015): 670.

6. W. Baade and F. Zwicky, "Cosmic Rays from Super-Novae." *Proceedings of the National Academy of Sciences of the United States of America* 20, no. 5 (1934): 259–263.

7. Baade and Zwicky, "Cosmic Rays from Super-Novae."

8. Two hundred Earth masses of gold are worth about \$275,000 trillion trillion, or 225 trillion times the world's gross domestic product (GDP). It is rather far away, however; traveling at the speed of light, it would take a quarter of a billion years to retrieve it.

17. IN THE BEGINNING

1. A. A. Penzias and R. W. Wilson, "A Measurement of Excess Antenna Temperature at 4080 Mc/s," *Astrophysical Journal* 142 (1965): 419–421

2. "Just 390,000 years" may sound odd, but we are comparing this to the age of the universe today at 13.81 billion years. For your seventy-three-year-old author, that's the equivalent of taking a picture 18 hours after birth, so it seems appropriate to call it a baby picture of the universe.

3. Blaise Pascal, *Pensées* (France, 1660).

EPILOGUE

1. E. R. Harrison, *Cosmology: The Science of the Universe* (Cambridge: Cambridge University Press, 1981).

Index

Page numbers of figures are *italicized*.